RURAL HIMALAYA ECOLOGY AND URBAN ECOSYSTEM

Himalaya Past and Present Series

RURAL HIMALAYA ECOLOGY AND URBAN ECOSYSTEM

Editor

K.S. Gulia

ISHA BOOKS
DELHI-110009

Rural Himalaya Ecology and Urban Ecosystem

Rs. 900

ISBN: 81-8205-426-5

Published in 2007 in India by
Isha Books
B-63, Ist Floor, New Gupta Colony,
Near DDA Market, Delhi-110009
Phone : 9212142040 E-mail : ishabooks@hotmail.com

Laser Typesetting at : Master Computers, Delhi
Printed at : Salasar Imaging System, Delhi

Contents

Introduction

India is fortunate to have the world's best mountain ranges. There are Aravallis, Nilgiris, Eastern and Western Ghats, and Vindhyachal and above all the Himalayas. Whereas the Himalayas are 40 to 50 millions years old, the others are more than 1500 million years old. The Himalayas extend from West to East, from the Pamir Kot in the northwest to the valley of the Brahmaputra River in the east. The range passes through Afghanistan, Pakistan, India, Jammu and Kashmir, Himachal Pradesh, Uttaranchal, Sikkim, Assam, Arunachal Pradesh, Manipur, Meghalaya, Mizoram, Nagaland, Tripura, West Bengal, Nepal, Bhutan, Bangladesh, Myanmar and China. Being the world's highest mountain chain, the Himalayas is characterized by a complex geologic structure, snowcapped peaks, large valley glaciers, deep river gorges and rich vegetation. 70 million people, covering 18 per cent of the geographical area and 6 per cent of population of India, inhabit the Himalayan mountainous belt. 'The mountain range with its richness of resources, has been contributing to the economic development of mainland India since times immemorial.

In the book, *Rural Himalaya Ecology and Urban Ecosystem*, we are discussing the Himalayas.

With population explosion, land use has undergone substantial changes in recent decades. This has in turn further affected water availability, both in terms of temporal as well as spatial distribution, which is expected to be highly vulnerable to anticipated climate changes. Because the melting season of snow coincides with the summer monsoon season, any intensification of the monsoon is likely to contribute to flood disasters in the Himalayan catchments areas. This has been discussed in Chapter One under the heading 'Main Issues and Concerns of Himalayan Ecology.'

Chapter Two is about History of Himalayan Ecology. The Himalayas are not of merely a geographical feature, i.e. a range of mountains. They epitomize a people's civilisational identity that goes back to the dawn of history. If these majestic mountains were not there, the rain clouds sweeping up from the Indian Ocean would have passed over the Indian sub-continent into Central Asia leaving it a burning desert. The compositions of the forests on this mighty range are effected by the combined effect of altitude, rainfall and latitude. The rainfall, mainly from the monsoons, decreases from the east to the west. These have been discussed in this chapter.

Chapter Three 'The People and Tribes in Rural and Urban Ecosystems' discusses the people of Ladakh, Bhutan, Sikkim, Nepal, Himachal Pradesh, and Uttranchal. From very early times, there have been migrations into the Himalayas and within it. Spiritual quest motivated a few to migrate there a small minority responded to the call of these mountains to test their own endurance and will power; pursuit of profit propelled others. Apart from the migrations that took place over the centuries, the tribes of India also require a deep look, Because the migrants were versatile and powerful, they kept on pushing back the original inhabitants of the land, till they receded very deep into the jungles. This Chapter elaborates these issues.

Chapter Four is 'Perspectives and Conservation Implications of Western Himalayas.' The Himalayan regions in India are characterized by all the problems associated usually with the mountainous environment of development countries, i.e. low availability of cultivable land, lack of diversity in economic activities, low accessibility, low productivity, meager infrastructure and inadequate employment opportunities. The problems are vast and complex as Himalayas itself is. These are interrelated and part of the Himalayan environment on the whole. The economy is basically agro-pastoral and the dependence on livestock is an integral part of the economy and ecosystem. This has been discussed in this Chapter.

Chapter Five is about Tibet and its Development. In the history of Tibet, one can find references to the development of religion — Buddhism or any particular philosophy — but hardly any reference to development. Broadly speaking the world development is accepted to mean an action or activity that causes a situation to change for the better. Many wondered how and why life in Tibet remained unchanged, unaffected by the new social economic and political ideas that transformed life in many other parts of the world. One could see that the Tibetans were always open to new ideas, adopting philosophy from India and many other cultural habits and customs from China and Mongolia. This Chapter discusses the lack of development in Tibet.

The forest tree canopy with its undergrowth of shrubs, perennial and annual herbs and the trunk laden with climber, epiphytic orchids, ferns, mosses and lichens. together with great stretches of bryophytes and ferns on the ground and on the rocks, constitutes the primary source of energy in the mountain ecosystem. Man in an effort to produce more food found it easier to extend his fields by clearing the adjoining forest rather than use improved agricultural technology, in doing so the forests overlooking valleys have been denuded over time. If one takes a trek anywhere in the mountains, one can notice the vegetation having undergone progressive deterioration, causing imbalances in the terrestrial ecosystem. This has happened all over the Himalayas. It has been discussed in Chapter Six 'Impact of Man on the Mountain Ecosystem.'

Chapter Seven is Kumaon: A Paradise in Peril Most people think of the Himalayas as ancient and strong, in fact, as a mighty protector of the country in the north. In reality, the Himalayan Mountains are geo-dynamically sensitive, youthful as compared to other mountain systems of the world and not particularly strong in a geological sense. The Himalayas largely consist of weak, deformed rocks which are severely stressed and easily susceptible to the onslaught of heavy rains, and the vibrations and disturbances caused by the movement of heavy traffic and human interference. Inadequately planned road construction, haphazard building constructions and continuing, deforestation aggravate the threat of landslides. All this has been discussed in detail in this Chapter.

Chapter Eight is Himalayan Rural and Urban Settlements. People living in the remote valleys of the Himalayas and in high-glaciated regions and alpine pastures are the people of nature. They and their ancestors have been living in the thick of nature for centuries. It is but natural that they show an uncanny skill in choosing correct sites for their settlements. The people from the plains who may have to choose sites for constructing new buildings may make mistakes in their choices. It is very important to have a correct appraisal of the environment to choose correct sites for new settlements and buildings. Otherwise it may lead to disaster.

Chapter Nine is on Himalayan Paryavaran. To understand the character of economic development and its impact on social-economic equilibrium in the hill region, demographic features must first be taken into account. Rapid increases in population impose tremendous pressure on land and other natural resources, disturbing the economic and ecological order in the hills. In this context, the Chapter takes a look at the origins of tea industry and the impact that it had on the Darjeeling area of the Himalayas. The British made the demarcation of land use in the Darjeeling hills in terms of tea and non-tea areas for purely historical reasons at the time of the leasing out of lands to the tea industry. The selective location of tea industry exerted a strong economic pull on migrant labour in Darjeeling. This has been discussed at length.

Chapter Ten discusses Development Projects and Ethno-religious Sentiments. Whenever the State takes up any development projects, friction with the populace of that particular area are bound to arise. The spectrum of such projects include very basic to highly sophisticated infrastructural development. Such projects often create undue confusion among the people. The issues raised by the critics of development projects may be as cognizable as conservation of environment and ecology, or as intangible as religious sentiment, tradition arid culture. In this Chapter the controversy over Rathong Chu dam in Sikkim has been discussed.

Sikkim has been inhabited by a number of communities — Lepchas, Bhutiyas and the Nepalis. The Bhutias came from Tibet in the 17th century and established their supremacy over the Lepchas who were the original inhabitants of Sikkim. The Nepalis who migrated en masse in the 19th century outnumbered them. When plainsmen from India immigrated, tension simmering under the surface came to the fore. Chapter Eleven tries to explore and elaborate how this relation in Sikkim has evolved over a long period of nearly 400 years and identifies the major phases and issues that have gone to shape the Sikkimese society as it is today.

1

Background Paper on Himalayan Ecology : Main Issues and Concerns

Introduction

India is fortunate to have the world's best mountain ranges. There are Aravallis, Nilgiris Eastern and Western Ghats and Vindhyachal. These are more than 1500 million years old. The Himalayas are comparatively young only 40-45 million years old. They traverse an arc for about 2500 km between the Indus and the Brahmaputra. The average width of this mountain range along the entire longitudinal extension ranges from 100 to 400 km. Upadhyay (1995) estimated that the Himalayas cover an area of 4.6 million sq. km. above 1500 m, 3.2 million sq. km. above 3000 m and 0.56 million sq. km. above 5400 m.

The Himalayas extend from west to east, from the Pamir Knot in the northwest to the valley of the Brahmaputra River in the east. The range passes through Afghanistan, Pakistan, India (Jammu and Kashmir, Himachal Pradesh, Uttaranchal, Sikkim, Assam, Arunachal Pradesh, Manipur, Meghalaya, Mizoram, Nagaland, Tripura and Darjeeling district of West Bengal), China, Nepal, Bhutan, Bangladesh and Myanmar.

About 140 million years ago some calamitous event took place. It was the break-up of the peninsular part of present India (The Deccan Plateau) from the super continent Gondwanaland. As a result of that Himalayas were born. The peninsular Indian Plate, having its original locale proximal to Madagascar and South Africa, started drifting northward about 15 cm per year, across what is now the Indian Ocean. Around 15 million years ago, it collided into the Asian continent near the present state of Assam in India. This crashing of the peninsular Plate against the southern edge of the Asian continental plate led to the rise of the Himalayas (Khoshoo, 1992). The Indian Plate continues to move northward at the rate of 2 cm per year. The Himalayas are still geologically active and structurally unstable. For this reason, earthquakes continue to be a frequent occurrence in the entire Himalayan region.

Geographical Structure

Being the worlds' highest mountain chain, the Himalayas is characterized by a complex geologic structure, snowcapped peaks, large valley glaciers, deep river gorges and rich vegetation. A complex interplay of climatic and geological processes, patterns of resource use and economic conditions have led to annual run-off of Indus, Ganga and Brahmaputra rivers is 208,494 and 510 km^3 per year, respectively.

No definite estimate can be made the quantity of water that flows down from the Himalayas into the rivers. Varying estimates of water resources in the Himalayan region have been made. Murthy (1978) estimated Himalayan water resources around 245 km^3 per year, Gupta (1983) and Kawosa (1988) estimated the total amount of water flowing from the Himalayas to the plains to be around 8643 km^3 per year. Bahadur (1998a) re-evaluated his earlier estimates of 200-500 km^3 per year as 400-800 km^3 per year as melt water contributions from the snow and glacier fields in the high mountain region. Despite these widely differing estimates of the water resource of the Himalayan region, the water output could be the highest from any single mountain range in the world (Stone, 1992).

With population explosion, land use has undergone substantial changes in recent decades. The hydrological characteristics of watersheds in the Himalayan region have undergone substantial changes this in turn has led to more frequent hydrological disasters, enhanced variability in rainfall and runoff, extensive reservoir sedimentation and pollution of lakes (Ives and Messerli, 1989).

This has further affected water availability, both in terms of temporal as well as spatial distribution, which is expected to be highly vulnerable to anticipated climate change (IPCC, 2001). Changes in the snowfall pattern have been observed in the Himalayas (Verghese and Iyer, 1993). These changes will have a marked impact on the monsoon regime, seasonal run-off and vegetation cover, including agriculture (IPCC, 2001). Because the melting season of snow coincides with the summer monsoon season, any intensification of the monsoon is likely to contribute to flood disasters in the Himalayan catchments (IPCC, 2001). Such impacts will be observed more in the Western Himalayas than in the Eastern Himalayas because of the higher contribution of snowmelt runoff in the west (Sharma, K.P. 1997).

Climate Change and Hydrology

The effect of climate change on hydrology of the region will be multifaceted: ranging from regional variations in precipitation characteristics, glacial shifts, mean run-off frequency and intensity of droughts and floods, soil moisture, water supplies for irrigation and hydroelectric generation. The Himalayas play a critical role in providing water to continental monsoon in Asia. Increased temperatures and enhanced seasonal variability in precipitation are expected to result in glacier recession and increasing danger from glacier lake outburst floods. A reduction in average flow of snow-fed rivers, together with increase in peak flows and sediment yields would have a major impact on hydropower generation, urban water supply and agriculture.

The paper is an attempt towards an analysis of the major ecological concerns, the underlying driving factors (both human and natural processes), socio-economic impacts and the responses in trying to cope with these, both at the micro/community level and at the national or international levels. We begin with a glimpse of the social and economic structure of the Himalayan region, which constitutes a critical basis for the ensuing analyses. This is followed by an overview of the state of the environment in the region and a detailed analysis of key environmental issues: water resources and their sharing, biodiversity loss, glacier recession, deforestation, land degradation and agriculture systems, natural disasters and solid waste management.

Socio-economic Background

The Himalayan mountainous belt is inhabited by 70 million people, covering 18% of the geographical area and 6 per cent of population of India. The mountain range, with its richness of resources, has been contributing to the economic development of mainland India since times immemorial. However, conventional parameters of development are not the best measures for understanding the social and economic status of the hill communities. The prevalence of traditional lifestyles, barter trade and common property resources tend to hide the level of poverty or prosperity of the highland people. The following discussion is structured around the key social and economic themes that have had ramifications for the state of environment and ecology of the entire Himalayan belt.

Cultural Diversity and Evolution of Traditional Practices

Through the last millennium, a continued process of observation and innovation led the communities of the Himalayas to evolve their economic activities and management strategies, primarily under conditions of steep topographies, inaccessibility and isolation, and need to maximize production while managing both risks and resources. Traditional strategies have been an outgrowth of both the advantages and disadvantages characterising survival in the small mountain communities.

Indigenous knowledge and practices generally emphasized low intensity of resource use in comparison with modern commercial lifestyles that have largely derived from ease of accessibility. The diversity of subsistence strategies comprised agricultural practices such as irrigation systems suitable for low rainfall conditions; shifting agriculture (jhum) in highlands of Sikkim and Arunachal Pradesh in the Eastern Himalayas, terracing of farmlands in Nepal and India in the middle Himalayas, medicinal uses of plants, management of forests, transportation and management of water and utilization of animal energy.

These strategies are illustrative of adaptive responses in coping with the needs and constraints of the highland environs. Recent research on this subject points toward a striking similarity in the pattern of social organization, land use and resource management practices amongst highland populations across the world, in spite of differences in regional, cultural and political history. It is interesting to note that ecological factors have had a remarkable impact on the evolution of social and cultural patterns and, hence, led to this convergence across subsistence type environments in the mountainous regions throughout the world (Mehta, 2001).

The cultural and linguistic diversity of the local communities has also been a dimension impacting ecological sustainability of the region. The existence of diversity, an asset during periods of prosperity and economic growth, could be a potential cause of widespread socio-economic instabilities, often generating hostility amongst various Himalayan communities, eventually undermining the concern for sustainable development (Stone, 1992).

Marginalization, Accessibility and Unsustainable Development

The evolution of socio-economic patterns of the Himalayas and its growing ecological fragility have to be viewed against the backdrop of issues of accessibility and geopolitical sensitivity. Historically, it was in the 1950s and the 60s that key events led to major socio-economic transformations. It is then that the newly independent nations of the Asian subcontinent that shared their boundaries in the Himalayas made efforts to improve accessibility to these remote mountain areas. The primary purpose was to ensure presence of state administration and institutions, often even for military and strategic reasons, in the formerly insulated and unapproachable border areas. Road building activities, air links and electronic communication, together with spread of market economy and policy interventions, have not merely impacted the socio-economic dynamics but also the rate of exploitation of natural resources in the region.

Economic development in the Himalayas in the last few decades has been characterized by felling of forests, enhanced exports of medicinal plants, development of hydropower projects, construction of water resource works, enhanced tourism, exploration and extraction of minerals, conversion of forested land to orchards and tea gardens, commercial horticulture and even establishment of cantonments and hill stations. The growth of industrial activity has been largely limited to the foothills. To this is added large-scale out-migration of male population in search of better job opportunities. This has affected the development of the region to a large extent.

Effects of Enhanced Accessibility

Whilst on the one hand enhanced accessibility led to political and economic integration of the mountain communities, on another it impacted the lifestyle and culture of the people there. However, this is not without its share of disbenefits. The political integration has thrust technological marginalization on them, by the entry of scientific experts from the mainland, equipped with systematic information about important mountain resources (forests, water, minerals), implementing programs and policies through both the centralised planning machinery and the forces of the market.

This economic amalgamation has often been said to reduce the autonomy of the local communities, with the risk of them being marginalized in the process of competing with the formal markets and the economies of the technologically modernized regions of the rest of the country. On the social front, migrant-urban encounters, tourism and exposure to images of urban life through popular media has led to spread of dominant values of the relatively rich peripheries and more powerful political centers in the plains, raising economic aspiration of the local hill communities and penetration of new values and modes of behaviour (Stone, 1992).

Poverty, Prosperity and Gender Differentiation

It is often difficult to quantify the poverty and prosperity in most parts of the Himalayan region on at least two accounts-the presence of barter trade and the predominance of common property resources (forests, rangelands, pastures). Increased commercial transactions in the mountain context has its pros and cons. With almost all the segments of the population impacted, the distribution of costs and benefits has been far from equal across communities and between men and women.

The spread of the "monetized economy" has its positive dimension in terms of providing access to jobs and cash incomes, employment opportunities, access to newer technologies and practices, etc. Increasing access to education has also been a positive outcome. In Chamoli district in the Garhwal region of Central Himlayas, there was a 52 per cent increase in literacy between 1961-71, 66 government intermediate colleges and 3 post-graduate colleges were also established (Jain, 1991). However, it was largely the men who benefited. As educated youth departed to seek employment away from their villages, the burden of subsistence work was transferred to the women and elderly.

Evidence from the Tehri Garwal districts pointed out that women, on their part, have been having less access to information on commercialized agriculture (agricultural extension workers, seed merchants, pricing systems, and credit institutions) to perform their agricultural work. This increasing dependence on market inputs and access to newer forms of information have been gradually divesting women's traditional role as the land manager to mere providers of labour (Mehta, 1994). The presence of markets and access to cash incomes implied that communities could do away with self-sufficiency. In fact, in many instances they seemed to have lost the capacity to meet their basic consumption requirements as a result of land degradation and diminishing agricultural returns.

Declining agricultural productivity and environmental degradation in parts of the Himalayas (and elsewhere) have also encouraged men and often even women to engage in short term migration or even seek employment in urban areas (Bourque and Warren, 1981; CWD, 1988; Schuler 1981) to fill the gap between subsistence production and consumption. Essentially it has been the women, children and elderly who are most affected by the fragmentation of social support structures.

Subsistence Needs versus Commercial Interests

The trade-off between meeting subsistence needs and commercial interests is engendering newer forms of social tensions. The privatization of the traditionally common property land along the "fruit-belt" of the Mussoorie Chamba road in Tehri Garhwal district has led to commoditization of local activities, use of migrant labor and changing working relationships. Women now have to traverse much longer distances to get to the forests; and grass, earlier freely available, has now to be bought at rather high prices.

In the villages of Bhagirathi valley, diminishing forest resources and declining agricultural base have acutely affected fodder supplies. The solution for social and economic development of the Himalayas lies in not merely enhancing access but at the same reducing the possibility of marginalization-through more prudent planning and utilization of natural resources and consideration for mountain specificities.

Overview of the State of the Environment

The Himalayan region is characterised not only by ecological fragility but also by a deep and historical geopolitical sensitivity (Stone, 1992). An ecosystem-based natural resource management approach is difficult to achieve as many countries share the resources of this mountain system. The Himalayas are full of natural wealth, both renewable and non-renewable resources. Amongst its non-renewable resources are deposits of boron, lead, lithium, coal, chromium, ores of iron, copper, tungsten, zinc and deposits of building material like limestone, dolomite and marble. These deposits occur across the length and breadth of the Himalayas cutting across international boundaries.

The Himalayas have substantial mineral wealth, the common image of resources there is that of water and forests (Stone, 1992). The Himalayas present a storehouse of bio-diversity, where flora and fauna vary extensively with climate diversity from one region to the other. If an attempt is made to divide the forests types based on the standard classification of tropical, sub-tropical, temperate and alpine, it becomes difficult to describe the rich diversity of the Himalayan forests.

Commercial Felling

However, extensive commercial felling of forest for timber has been reported in the region, especially in the last few decades when urban centres began to grow in areas near the forests. While the forests, farmlands and grasslands are of extreme importance for the agripastoral economy within the Himalayan region, the other renewable resource, water, has always been crucial for thickly populated plains further down in the south and east. The Himalayas has one of the highest hydropower potentials of the world which includes three of the mightiest rivers of the world, i.e., Indus,

Ganga and Brahmaputra. Besides, water resources, vast extended grasslands and dense forests in some highland parts and in the foothills of the Himalayas have certain specific advantages. These areas have a tremendous potential for tourism.

There is a drawback. Extreme climate variations and inaccessible terrain make it difficult to exploit and utilise all the natural resources of the great Himalayas. Climatic extremes like cloudbursts, hailstorms and earthquakes play a critical role as an environmental constraint. Global warming is affecting ice and glacier cover in the region. Furthermore, due to inaccessibility of the region, it is difficult to quantify and assess the damage caused by these natural events. The environmental constraints against the exploitation of natural resource also include other natural phenomena like mass wasting, high seismic activity, landslides, glacial lake outburst floods, erosion and sedimentation. Soil erosion is also one of the major environmental constraints, which results in frequent flooding in the plains downstream and damage to agriculture, life and infrastructure.

Hazard by Grazing

The livestock population in the region has also increased during the recent past and problem of grazing in the high altitudes has acquired serious dimension, as a result the vegetative cover is decreasing which has resulted in the loss of top soil due to excessive erosion. The relatively soft rocks and favourable climatic conditions are the factors behind quick weathering followed catchment in the Western Himalayas (the Chenab, a tributary of the Indus) show that the average snowmelt and glacier-melt contribution to the annual flow of water is 49.1%; a significant proportion of run-off is derived from snow in the dry season, when water demand is highest (Singh et al., 1997).

If the rainfall pattern in the Himalayas changes due to climate change, the impact could be felt in downstream countries such as Bangladesh. The water and agriculture sectors are likely to be most sensitive to climate change-induced impacts in Himalayas. Catchments in Nepal supply about 70 per cent of the dry-season flow of the Ganges river, and the tributaries of the Brahmaputra river originating in Bhutan supply around 15 per cent of the total annual flow of that river. If climate change disrupts these resources and alters mountain hydrological regimes, the effect will be felt not only in the mountain core of Tropical Asia but also downstream, in countries that are dependent on this water resource.

Sharing of Water by Countries of the Region

Waters in the South Asian region acquire strategic importance as many countries share these resources. There have been water disputes between India and Nepal, Pakistan and Bangladesh over sharing of waters and treaties have been signed from time to time in an attempt to solve the problem of water sharing in the region. The most significant bilateral treaties in this respect are the Indus Waters Treaty (1960), the Ganges Water Sharing Agreement (1996) and the Mahakali Treaty (1996). A discussion of these treaties is presented below.

Water Treaties

The India-Bangladesh treaty on sharing of waters of the Ganga river and the India-Nepal treaty on sharing of the waters of the Mahakali river are intended to bring to an end to long running differences between India and her neighbors over the entitlement of water flows following the construction of barrages on these rivers. The treaties establish long term water discharge regimes of 30 and 75 years respectively, focussing on the utilization of waters rather than their conservation.

India-Pakistan

The Indus River originates at 17,000 feet above the sea level in Tibet, traverses 1800 miles through the Himalayas in Jammu and Kashmir and Pakistan before joining the Arabian sea. The flow of Indus fluctuates seasonally, with melting of Himalayan glaciers accounting for almost 90 per cent of the water in the upper Indus river basin (Rizvi, 2001). The division of the Indus river basin between India and Pakistan following the partition of the countries in 1947 resulted in a break out of disputes over water sharing.

Indus Water Treaty : The Indus water treaty between the two countries was signed in 1960 and divides the catchment area between the two countries. The treaty is regarded as one of the few successful settlements of a trans-boundary water basin conflict. Under the treaty India was granted the three eastern tributaries (Ravi, Beas and Sutlej) and Pakistan assumed the flow of three western rivers (Indus, Jhelum and Chenab). Other provisions of the treaty have enabled construction of two dams in Pakistan (Mangla and Tarbela) and created new link canals and barrages to develop and sustain agricultural activities.

India-Bangladesh

In 1960, negotiations started between India and Bangladesh on the sharing of the waters at Farakka. An agreement was signed in 1977 for a period of five years, by which 63% of the Ganges water was to be allocated to Bangladesh in dry season (1 January to 31 May) (Verghese & Iyer, 1993). This agreement was followed by MoUs (Memorandums of Understanding) in 1982 and 1985. Finally, in December 1996 India and Bangladesh signed the Ganges Water Sharing Agreement, a 30-year treaty. The Indo-Bangla treaty makes reference to a number of guiding principles. It aims to make "optimum utilization" of the waters of the region, bringing a "fair and just" solution to the Farakka water problem. The Indo-Bangla treaty establishes a Joint River Commission with the mandate of monitoring daily flows, submitting data and implementing arrangements under the treaty.

India-Nepal

The Mahakali River (known as the Sharda River in India) has formed the border between India and Nepal since 1816, and has been the subject of tension between the two countries over sharing of its resources. In 1996, India and Nepal signed the Mahakali Treaty, for a period of 75 years. The treaty settles Nepal's entitlement to waters from the existing Sarada barrage and authorizes (without prejudice to Nepal's sovereign rights over that land) India's prior construction of the part of the recently constructed Tanakpur barrage, which occupies 2.9 hectares of Nepalese territory. In return Nepal would receive an agreed supply of hydropower. The treaty also creates the framework for the construction of the integrated Pancheshwar multipurpose project on the Mahakali River.

Mahakali River Commission : A joint commission, the Mahakali River Commission has been established by the India-Nepal treaty, which makes recommendations for the conservation and utilization of water, evaluation of projects and examination of differences between the parties concerning interpretation and application of the treaty (Sands, 1996). The construction of the Pancheshwar dam on the Mahakah River has raised some protest and opposition in Nepal, as the site falls in the seismically volatile Himalayan region. If the dam breaks due to an earthquake, a major catastrophe could result that would threaten the lives of millions of people on both sides of the Indo-Nepal border.

Institutional Responses

As many countries in South Asia share the water resources of the Himalayas, regional cooperation is essential for sustainable utilization of these resources. Existing efforts at cooperation are bilateral in nature and there should be a move towards greater regional cooperation in water resources development and management, to provide easy access to relevant data and information, greater dialogue and information exchange among research institutions, experts and policy makers.

United Nations Educational, Scientific and Cultural Organisation's (UNESCO) recently formed FRIEND (Flow Regimes from International Experimental and Network Data) is one such effort, which envisages regional cooperation in the exchange of hydrological data in Hindu Kush Himalayan region. The International Centre for Integrated Mountain Development (ICIMOD), at Kathmandu, has carried out programs on water in association with the International Hydrological Program (IHP) of UNESCO and the Operational Hydrological program (OHP) of the World Meteorological Organisation (WMO). The institute has launched the Hindu Kush Himalayas (HKH)-FRIEND project, which has been endorsed by six of the eight HKH countries.

Coordinated and intensified regional efforts could bring forth solutions to water shortages for irrigation and drinking, generation and sharing of hydropower and conjunctive utilization and management of water resources. There is a growing consensus that regional cooperation in the use of great rivers of South Asia, particularly the Ganga, Brahmaputra and Indus is essential to solving some of the regions environmental problems. Regional cooperation in water sharing would be a step towards increasing environmental and regional security in the region.

Glaciers

Glaciers are moving bodies of ice and snow, which are normally present above the snow line. Glaciers and ice sheets are hundreds to more than one thousand meters thick and change significantly only over decades. On these longer time scales, they can influence atmospheric circulation and global sea levels. Glaciers play an important role in maintaining ecosystem stability as they act as buffers and regulate the runoff water supply from high mountains to the plains during both dry and wet spells.

The Himalayan glaciers account for about 70 per cent of the world's non-polar glaciers and affect the lives of millions of people in several countries: China, India, Pakistan, Afghanistan, Nepal and Bangladesh. Their runoff feeds two of the oldest rivers in the world, the Indus and the Ganges, whose tributaries carry precious water for 500 million people on the northern Indian plains. Most of the glaciers in the Himalayas are of a summer-accumulation type, that is major accumulation and ablation take place simultaneously during summer (Fujita. et. al, 1997).

The Longest Glaciers

The glaciers of the Himalayas include some of the longest outside the Polar Regions and reached their largest extent during the end of the last ice age (more than 20,000 years ago). The evidence of these large ice masses can be seen in 'J' shaped valleys, which characterize much of the higher Himalayas. On the basis of their mode of occurrence and dimensions, glaciers have broadly been classified into three categories: valley glaciers, piedmont glaciers and continental glaciers. Himalayan glaciers fall in the category of valley glaciers.

It has been estimated that an area of about 32,000 sq. km is under permanent cover of ice and snow in the Himalayas (Negi, 1991). This amounts to about 179/6 of the total geographical area of the Himalayas. Higher concentration of glaciers in the Himalayas lie in the regions with the highest mountain peaks, that is, Nanga Parbat, Nun Kun, Kinner Kailash, Nanda Devi, Nanda Kot, Annapurna, Mt. Everest, Makalu and Kanchanjunga. There are a number of small, medium and large size glaciers in the Himalayan ranges with typical landform features. Some of the famous and important ones include Baltoro glacier, Gangotari glacier, Gasherbrum glacier, Siachen glacier, Kanchanjunga glacier and Hispar glacier. Of these, the Siachen glacier is the most well known, on account of its strategic significance in the South Asian region (Box 1).

The Siachen Gaicier

The Siachen glacier lies in the extreme north-central part of Jammu and Kashmir near the border of India and Tibet. It has a length of about 72 km and is the largest glacier in the world outside the polar regions. It derives its name from "Sia" meaning roses in Baltic. With the presence of the military in Siachen since 1984, it has become famous for being the highest battle ground, the highest helipad and the highest drop zone. The glacier, which is the source of the river Indus, is facing the threat of degradation as it has become a dump yard of cans, drums, tetrapacks, aluminium foils and medical waste.

Environmentalists and mountaineers have expressed their concern over the degradation of the glacier and proposals to turn Siachen into a peace park have been made.

Glacial Recession

Glaciers are dynamic in nature; they grow and shrink in response to changing climate. During the Pleistocene era (2 million years ago) glaciers occupied about 30 per cent of the total area of the earth as against 10 per cent at present (Bahadur, 1998b). The United Nation's Intergovernmental Panel on Climate Change has stated that thinning of glaciers since the mid-19th century has been obvious and pervasive in many parts of the world. The reduction in ice cover during the last century, across the globe, especially in mountain glaciers is seen as evidence of climate change. Global temperatures have risen by 0-03 to o.o60 C since the last century, as a result of which mountain glaciers have thinned, lost mass and retreated. This retreat is consistent with an observed alpine warming (Oerlemans, 1994).

The melt waters released by glaciers are a major source of erosion in barren areas having little vegetative cover. While receding, glaciers leave behind unconsolidated debris called terminal moraines, which may become the source material for landslides in future. Scientists have expressed concern that if glaciers continue to shrink, landslides, avalanches and floods will increase.

Glacier Recession

A study conducted by thb International Commission on Snow and Ice (ICSI) has observed that Himalayan glaciers are receding faster than in any other part of the world and if the present rate continues, the likelihood of them disappearing by the year 2035 is very high. The Gangotri glacier, one of the major and important glaciers in the Himalayas, was 25 km long when measured in the 1930s and has now shrunk to about 20 km (Hasnain, 1999).

The Dokriani Bamak glacier (3 miles long) in the Himalayas has been one of the most studied in the world, and since 1990 it has receded half a mile (Gergan et. al, 1999). Research indicates that glaciers in the Himalayas are retreating at an average rate of 30 m a year, compared with earlier rates of 18 m a year between 1935 and 1999, and 7 m a year between 1842 and 1935 (www.worldwatch.org).

Glacial Lake Outburst Floods

The widespread glacial retreat in the Himalayas has resulted in the formation of many precarious glacier or moraine-dammed lakes. These glacial lakes are unstable and most of them are potentially susceptible to sudden discharge of large volumes of water and debris which causes floods downstream, commonly known as Glacial Lake Outburst Floods (GLOF). In the high altitude region (>3500 m) energy releases in the form of GLOF could result in heavy loss of life and property, and have trans-boundary impacts. In recent years several GLOF have been reported in India, Pakistan, Nepal, Bhutan and Tibet-China resulting in widespread damage to life and property. Five GLOF events occurred in Nepal from 1977 to 1998. The latest GLOF in Bhutan, which occurred in 1994, was triggered by partial breaching of the Lugge Tsho glacial lake.

The other two notable GLOF events in Bhutan occurred in 1957 and 1969. In the Tibetan area of the Himalayas, as many as 12 GLOF events have been recorded since 1935 (Mool, 2001). The 1985 Dig Tsho Glacial Lake Outburst Flood in Nepal's Langmoche valley caused severe loss to life, property and infrastructure, drew the attention of scientists, planners and policy makers and led to study/ field investigation of several other glacial lakes in the region. An inventory and field investigation of glacial lakes in Nepal, China and Bhutan commenced in the 1980s Nepal-China joint investigation results show that in the Tibetan region, there are 24 potentially dangerous glacial lakes in Pumqu (Arun) river basin and lo in Pioqu (Bhote-Sun Koshi) river basin (LIGG/WECS/NEA, 1988).

A recent inventory carried out by ICIMOD and the United Nations Environment Programme/ Environmental Assessment Programme for Asia and the Pacific (UNEP/ EAPAP) shows that there are 26 potential dangerous glacial lakes in Nepal (ICIMOD and UNEP/EAP-AP, 2000). A similar inventory carried out for Bhutan shows that there are 22 potentially dangerous glacial lakes. What is necessary at this stage is detailed inventory of glaciers and glacial lakes of the entire Himalayan region to monitor and mitigate the potential risk and impact of GLOF.

Biodiversity Loss

The Himalayan mountain system represents one of the richest natural heritage sites in the world. A conglomeration of topographical and climatological regimes is manifested in a remarkable assemblage of biodiversity-both plants and animals. One-tenth of the world's known species of higher altitude plants and animals occur in the Himalayas (IPCC, 2001). Rodgers

Table 1.1 : Biotic Provinces of the Indian Himalaya

Biogeographic provinces	*Political boundaries*	*Area (sq. km.)*	*Total area (%)*
Trans Himalayas	Jammu & Kashmir Himachal Pradesh	186,200	44.07
North west Himalayas	Jammu & Kashmir Himachal Pradesh	69,000	16.33
West Himalayas	Hill districts of Uttar Pradesh* (Kumaun and Garhwal)	72,000	17.04
Central Himalayas	Sikkim & North Bengal (West Bengal)	12,300	02.91
Eastern Himalayas	Arunachal Pradesh	83,000	19.64

* The hill districts of Uttar Pradesh are now part of the newly formed state of Uttaranchal.

Source: After Rodgers & Pawar (1988) adapted from Dhar U. (1996)

(1988) divided the Himalaya into five biotic provinces under two biogeographical zones: Boreal and Indo-Malayan Table 1.1. The west Himalayan biotic province supports cold and drought resistant vegetation dominated by conifers, legumes, grasses and composites. The East Himalayan flanks harbour a wet humid subtropical vegetation rich in magnolias, oaks, laurels, terminalis, rhododendrons, epiphytes, orchids and ferns.

These biogeographic provinces reflect a variety of ecosystems including the famous cold desert. Sprawled over a vast area north of the Himalayan ranges, the cold desert is an ecosystem of exceptionally low temperatures down to-75°C and rainfall 500-800 mm annually. The desert forms a plateau at the height of 4500 to 6000 ft and is encompassed by the trans-Himalayan bio-geographic Zone described by Rodgers and Pawar (1988). This zone extends into the Tibetan Plateau, to cover an area of 2.6 million sq. km, from which originates the great river system of Indus, Sutlej, Brahmaputra and Yangtze.

Vast Vegetation

The Himalayan region supports a vast and luxuriant vegetation ranging from tropical and subtropical to alpine (including alpine scrubs of the cold desert) with a rich gene pool of cultivated plants. Owing to a great variety of physiographic and phytoclimatic conditions, the Himalayas foster almost all types of vegetation from humid tropical evergreen to moist mixed deciduous sal forests, marshes, swamps, mixed deciduous forests, subtropical pine forests, broad leafed temperate forests, moist temperate broad leafed deciduous forest, temperate conifer forests, subalpine and alpine vegetation, alpine meadows and alpine scrubs. Table 1.2 presents an overview of plant diversity in the Himalayan region. The Himalayan region also possesses the richest diversity of medicinal and aromatic plants in the country (Arora, 1993).

Table 1.2 : Plant Diversity in the Himalayan Region

Total number of species	*Himalayan region*	*India*	*World*
Angiosperms	8000 (3200)	17000 (5400)	250000
Gymnosperms	44(7)	54(8)	600
Pteridophytes	600(150)	1022(200)	12000
Bryopthytes			
Liverworts	500(115)	843(166)	8500
Mosses	1237(450)	2000(820)	8000
Lichens	1159(130)	1948(423)	20000
Fungi	6900(1890)	13000(3000)	120000

Note: Figures in parenthesis represent the number of taxa considered endemic to Himalayan/Indian region.

Source: Singh and Hajra, 1996.

Similarly, fauna in this region presents one of the richest assemblages in the Indian subcontinent. Many species such as the Snow Leopard, the Himalayan Brown Bear, Red Panda, Himalayan Lynx, Kashmir stag, Himalayan Musk Deer, Yak, Himalayan Ibex, Himalayan Thar and the Himalayan Bearded Vulture are unique to the region.

Species Diversity

Species diversity both amongst invertebrates and vertebrates appears to be very significant and includes many primitive, new and evolving forms. The vertebrate fauna in the Himalaya provides a high degree of diversity at species level. Out of the total mammalian species, 65 per cent are recorded in the Himalaya; 50 per cent of the total bird species occur in the region and likewise 35 per cent reptiles, 36 per cent amphibians, and 17 per cent fishes could be documented from the mountain ecosystem establishing the status of the area as a center of origin and evolution of biotic forms (Ghosh, 1997). Further, 29 Out of 428 species of Reptiles from India, 35 species of amphibia

(Out of 200) and 36 species of freshwater fishes (out of 1300) are endemic to the Himalayan region (Ghosh, 1997).

Threat to Bio-diversity

The Himalayan region is a prime center of diversity in the sub-continent and also in the entire world. However, the rich diversity of Himalayas is now severely threatened and many species have become endangered on account of a variety of physical, biotic and strategic factors unique to the region. The loss in biodiversity has been largely on account of habitat losses which in turn have happened due to deforestation, overexploitation of resources, wetland drainage and filling, human settlements/encroachments, population growth, and species introduction. The rich endemic flora of the Himalayan region faces the threat of extinction due to degradation of forests and natural habitats for expansion of agriculture, river valley projects, industrial and urban development (MoEF 1999).

One of the big challenges is the natural resource dependence of the native population of the region, who are primarily agro-pastoral. The bulk of the region has dependence on livestock. The potential impacts of excessive grazing include depletion of the scarce forage for wildlife, habitat degradation, disease transfer and reduction in the breeding performance of both wildlife and domestic stock. This leads to livestock depredation by wild carnivores, which is a serious conservation issue in the Himalayan region.

Forest Fires

Other factors for loss of biodiversity include forest fires (which affect regeneration) and natural calamities. Habitat destruction and fragmentation has resulted in the vulnerability of species to inbreeding depression and high infant mortality. The construction of mines, roads, dams and other infrastructure has also resulted in loss and degradation of habitats.

The region is transversed by many international boundaries. Poaching and illegal trade in wildlife are major threats to species survival in the Himalayas. It provides transit routes for illegal wildlife trade. Of the various Himalayan species, plants and plant products are among the most heavily traded. Among fauna, Musk Deer, Himalayan Black bear, butterflies and Tibetan antelopes for Shahtoosh are sought for their high value in the international market. Further, due to the sensitive nature of many of the international boundaries, biodiversity conservation in the region becomes difficult to administer. It is an open field for poaching. As a result of these processes, many endemic species of the Himalayan region such as the Himalayan Lynx, Snow Leopard, Himalayan Thar, the Yak and avian species such as the Himalayan bearded Baza and the Mountain Quail are now endangered. Approximately 10 per cent of known species in the Himalayas were listed as threatened in 1995, and the number of species on the verge of extinction has increased since then (IPCC, 2001).

Institutional Responses

Environmental protection is enshrined in the Indian Constitution, which enjoins that the state shall endeavour to protect and improve the environment and safeguard forests and wildlife in the country. Government action towards conserving biodiversity has taken the form of creation of protected area networks (biosphere reserves, national parks and wildlife sanctuaries) and enforcing legislation for the protection of plant and animal species. The protected area networks in the Himalayas cover about 10 per cent of its geographical area, which is also the norm recommended by The World Conservation Union (IUCN). There are 31 National Parks and 136 Wildlife Sanctuaries in the region Table 1.3. Biosphere reserves are located at

Dehang Debang in Arunachal Pradesh, Khangchendzonga in Sikkim and Nanda Devi in the state of Uttaranchal. The Nanda Devi reserve with its rich biodiversity and extending over an area of more than 2000 sq. km., has been declared as a World Heritage site.

Table 1.3 : Protected Areas in the Himalayan Region in India

State	*National Parks*		*Wildlife Sanctuaries*		*Total*
	Number	*Area (sq.km.)*	*Number*	*Area (sq.km)*	*(sq. km.)*
Arunachal Pradesh	2	2468	10	7114	9583
Assam	3	1174	13	940	2114
Himachal Pradesh	2	1429	32	5737	7166
Jammu and Kashmir	4	4650	16	10172	14822
Manipur	2	82	1	185	267
Meghalaya	2	267	3	34	302
Mizoram	2	250	4	634	884
Nagaland	1	202	3	24	226
Sikkim	1	1784	5	265	2049
Tripura	0	0	4	604	604
Uttar Pradesh	7	5411	29	7595	13005
West Bengal	5	1693	16	1103	2796
Himalayan States total	31	19410	136	34408	53818
All India Total	**87**	**40362**	**485**	**115374**	**156006**

Source: FSI 2000.

Legal Framework

An extensive legal framework for the protection of environment and wildlife exists in India. Main legislations include the Indian Forest Act 1927, Wildlife (Protection) Act 1972, Forest Conservation Act 1980 and the Environment (Protection) Act 1986. The Wildlife (Protection) Act is the single most significant statute on wildlife conservation in India. More recently, specific national plans for the protection of biodiversity have been developed and introduced in the Parliament. The Biodiversity Bill 2000 aims to provide for conservation of biological diversity, sustainable use of its components and equitable sharing of benefits. A National Strategy and Action Plan on Biodiversity has also been prepared by the Ministry of Environment and Forests.

India is also a signatory to multilateral environment agreements such as the Convention on International Trade in Endangered Species of Wild Fauna and Flora (CITES) 1973 and the Convention on Biological Diversity (CBD) 1992 for the protection and conservation of biodiversity.

Deforestation

Forests are repositories of biodiversity and have been used extensively for thousands of years as sources of food, fuelwood, fodder and other biomass by the Himalayan people, thus playing a key role in the agri-pastoral economy in the region (Stone, 1992). Forests perform invaluable ecological services such as controlling soil erosion: trees slow down run-off and control soil erosion and landslides by retaining water in their root systems and by arresting the velocity of rain water through their canopy cover (TERI 1992).

Forests also act as sinks for storing carbon dioxide emissions. The climatic and rainfall variations in the Himalayan region have resulted in a wide variety of forest ecosystems, ranging from tropical forests rich in plant diversity in the Eastern Himalayas to alpine forests in the cold and dry western Himalayan region. The western Himalayan region in India extends from Kashmir to Kumaon. The temperate zone of this region is rich in forests of spruce, fir, cedrus, chir, pine, other conifers and broadleaved temperate trees. The alpine zone extends from the upper limit of the temperate zone to about 4,750 m elevation, with the characteristic trees of the region being silver fir, birch, juniper and dwarf willows (MoEF 1999).

Table 1.4 : Extent of Forest Cover in the Himalayan States (sq. km.)

State	*Geog. Area*	*Dense Forest*	*Open Forest*	*Total Forest cover*	*% of All-India forest cover*	*% of geog. area*
Assam (hill districts)	17484	8039	5218	13257	2.08	75.8
Arunachal Pradesh	83743	57756	11091	68847	10.80	82.21
Himachal Pradesh	55673	9120	3962	13082	2.05	23.5
Jammu and Kashmir	222235	11019	9422	20441	3.21	9.2
Manipur	22327	5936	11448	17384	2.73	77.86
Meghalaya	22429	5925	9708	15633	2.45	69.7
Mizoram	21081	3786	14552	18338	2.88	86.99
Nagaland	16579	5137	9027	14164	2.22	85.43
Sikkim	7096	2363	755	3118	0.49	43.94
Tripura	10486	2228	3517	5745	0.90	54.79
Uttar Pradesh (hill districts)	51134	17584	5091	22675	3.56	44.34
West Bengal (Darjeeling district)	3149	1096	359	1455	0.23	46.20
Himalayan states Total	533416	129989	84150	214139	33.60	40.1
All-India Total	**3287263**	**377358**	**255064**	**637293**	**100.00**	**19.39**

Notes:
1. Assam hill districts: Karbaianglong and North Cachar hills.
2. The hill districts of Uttar Pradesh are now part of the state of Uttaranchal and comprise: Almora, Chamoli, Dehradun, Garhwal, Nainital, Pithora-garh, Tehri-Garhwal and Uttarkashi.

Source: FSI 2000.

The Eastern Himalayan region extending eastward from Sikkim has forests of oak, laurel, maple, alder and birch, juniper, conifers and dwarf willows. The richness of the Himalayan forest declines from east to west and from south to north (Stone, 1992).

The Himalayan states together account for one-third of the total forest cover in the country, with Arunachal Pradesh alone contributing 10.8 per cent of the total forest cover Table 1.4. Forests cover 40 per cent of the geographical area in the Himalayan region and between the two assessment periods of 1997 and 1999, forest cover in the states of Arunachal Pradesh, Himachal Pradesh, Jammu and Kashmir, Tripura, Uttar Pradesh and West Bengal increased (FSI 2000). A decline in forest cover has been witnessed in the North-eastern states of Assam, Manipur, Meghalaya, Mizoram, Nagaland and in Sikkim.

Forest Cover

Despite this decline in forest cover in these states in the past two years, forest cover as a proportion of total geographical area is well above the 66 per cent recommended by the National Forest Policy of 1988 (except in Sikkim). The recent increases in forest cover in some states however do not reveal the extent of pressure on these resources in the Himalayan region, particularly in states such as Himachal Pradesh and Jammu and Kashmir, where forests occupy a relatively small proportion of total area.

Deforestation

Forest areas in the proximity of population centers and villages are reported to be degrading faster due to collection of fuelwood and cattle grazing, as compared to forests away from population centres and located in inaccessible areas (FSI 2000). Deforestation and degradation has been caused by commercial logging; clearing of forestland for settlements and agriculture; excessive exploitation of forests for fuelwood and food; and overgrazing. The denudation of fir and spruce forests in Himachal Pradesh and Jammu and Kashmir for making package cartons for apples is an example of the commercial exploitation of forests taking place in the region (TERI 1992).

Fuelwood constitutes the dominant source of energy in the region and population growth has contributed to increased exploitation of forests for meeting growing energy demands. Limestone quarrying and the construction of dams added to the loss of forestland. Increases in animal population have contributed to degradation of grasslands and pastures. The absence of organised forestry, the demand for wood for constructing sleepers for the railways and military demand during the two world wars also led to deforestation in the Himalayas (Singh, 1992).

Shift in Land Use Pattern

Land use patterns such as shifting cultivation (*Jhum*) in the growth-eastern region have contributed to stress on forest resources. An assessment by the Forest Survey of India indicates that 1.73 million hectares of land has been affected by shifting cultivation in this region during the period 1987-1997 (FSI, 2000). Nagaland and Mizoram together account for 65 per cent of the area under shifting cultivation in this region. Due to degradation of forest cover and pastures, farmlands which are dependent on them for nutrients and organic matter also deteriorate. This loss of extent and productivity of forests and pasturelands implies a threat to the well-being of rural communities who are dependent on the agri-pastoral economy for their livelihood. The impact of deforestation is felt most by local communities, particularly women and children, who as a result of receding forest cover have to travel increasingly longer distances and spend more time on fetching fuelwood and fodder, contributing to severe hardship.

The loss and degradation of forests affect the ecological services performed by them. The consequences of this could be the conversion of sinks for carbon dioxide emissions to sources of release of vegetative carbon into the atmosphere; changes in forest composition and possible disappearance of some forest types, increased problem of landslides and soil erosion (Singh, 1992). Hydrological imbalances currently observed in the Himalayas are considered to be linked to the loss in vegetation cover (Farooquee and Rawat, 1997).

Institutional Responses

Local communities in the Himalayan region have played a very important role in forest protection. The Chipko movement, the first organised people's environmental movement in India, had significant success in protecting the sensitive forest eco-system in the Himalayas (Human and Pattanaik, 2000). The movement began in the early 1970s when, led by Chandi Prasad Bhatt and Sunderlal Bahuguna, people in the hills of northern Uttar Pradesh (now the state of Uttaranchal) protested against the Government's allotment of forest land in the Alakananda valley for commercial logging. The unique method adopted by the villagers was that of hugging the trees to physically prevent their logging.

This drew the attention of the Uttar Pradesh state government who instituted a committee to investigate the matter and concluded that it was because of widespread deforestation in the catchment of the Alakananda river that disastrous flash floods had occurred there in 1970 (Human and Pattanaik, 2000). As a result, a complete ban was placed on all commercial felling in the upper catchment of the Narmada river and its tributaries for ten years. In the next decade the Chipko movement spread from the Garhwal hills to other parts of the Uttar Pradesh Himalayas and protests against deforestation were organised through mass mobilisation and demonstrations. In 1981 a demand was made for a total ban on tree felling in the Himalayas above an altitude of 1000 m. In response, the government constituted a committee to prepare a comprehensive forest policy for the Himalayas (Human and Pattanaik, 2000). Subsequently, the government imposed a moratorium on all commercial felling throughout the Himalaya hill zone for the next 15 years. The Chipko movement is thus a landmark in the environmental movement in India.

Government efforts at forest conservation have included the development and implementation of legislation, programmes and policies. An overview of forestry sector legislation and main policies followed by the government for promoting community participation in the management and protection of forests is discussed in this chapter.

Indian Forest Act 1927

The Indian Forest Act 1927 is the principal legislation for the management and protection of public forests in the states. Thereafter, specific and revised laws have been passed relating to specific fields of forestry and new forest policies have been issued (MoEF, 1999). The Forest (Conservation) Act 1980 was enacted with a view to check the diversion of forest land for non-forest uses. As a result, approval of the Central government is required before diverting forest land for non-forestry purposes.

In addition, the diversion has to be accompanied by compensatory afforestation in an equivalent area of non-forest land or double the area in degraded forest land. In addition to these basic legislation in the forestry sector, there exist complementary legislation relating to forests in the Mines Act (1952), the Wildlife (Protection) Act 1972 and the Environment (Protection) Act 1986. This framework of legislation seeks to set priorities for maintaining existing forestlands, setting aside land as protected areas, environmental protection and conservation and meeting the needs of the rural poor and tribal people under traditional rights and concessions (MoEF, 1999).

Promotion of local community participation has been a highlight of recent developments in forest sector policies. The Joint Forest Management (JFM) being the mechanism for implementing this policy in the states. The programme was adopted in 1990 to encourage people's involvement in forest management, regeneration and development of degraded forests and to provide fuelwood, fodder, non-wood forest products and timber to people living in adjoining areas. Since then all the Himalayan states (except Manipur and Meghalaya) have issued notifications entitling communities to sharing of user rights in forests and involving them in the protection and development of forest areas. A consolidation of government efforts towards promoting greater local involvement in forest management was provided by the Constitutions 73rd Amendment Act, 1992. This assigned forestry functions including conservation, management and sustainable development of forests to village panchayats.

National Forestry Action Programme

The most recent development in forest policy in India is the formulation of a National Forestry Action Programme in 1999, which identified five areas of action for the sustainable development of forests: protecting existing forest resources, improving forest productivity, reducing demand for forest products, strengthening policy and institutional framework and expanding forest area. Rehabilitation and increase in productivity of degraded forests and an increase in the area under forest and tree cover to 33% of total land area is proposed.

At the international level, since the early 1980s, there has been extensive concern on the loss of forests around the world, particularly tropical forests. In the 1990s, prior to the United Nations Conference on Environment and Development (UNCED), this concern was expressed in the form of a demand by developed countries for an international treaty to curb deforestation. However, agreement on an international treaty could not be reached as developing countries raised concerns about the focus on tropical forests, and issues of sovereignty and control over natural resources (Fletcher, 1995).

Action on Rio Summit

As a result, the Rio summit concluded with the formulation of a "Non-legally binding authoritative statement of principles for a global consensus on the management, conservation and sustainable development of all types of forests", more popularly known as the Rio Forest Principles. The second document that emerged from the summit was the inclusion of an action plan to be followed by countries for "Combating Deforestation", in Agenda 21, the road map to sustainable development prepared at the summit. The mechanism through which the Rio Forest Principles and the recommendations for combating deforestation in Agenda 21 are to be implemented is the Commission on Sustainable Development (CSD). In 1995, the CSD established the Intergovernmental Panel on Forests (IPF) to implement the UNCED decisions related to forests at the national and international level.

The IPF has made over a hundred proposals for action on a number of issues related to sustainable forest management (United Nations, 2001). To carry forward the inter-governmental policy dialogue on forests, the Intergovernmental Forum on Forests (IFF) was established in 1997 under the CSD. Its mandate comprised promoting and facilitating the implementation of the proposals for action of the IPF and working towards a consensus on international arrangements and mechanisms for the sustainable development of all forests, including, for example, a legally binding instrument. At the fourth and final session of the IFF in February 2000, the United Nations Forum on Forests (UNFF) was established as a new non-legally binding body to facilitate and promote the implementation of proposals for action emanating from the IPF and the IFF. The most recently established international partnership on forests, is the Collaborative Partnership on Forests (CPF), formed in April 2001. The initial membership comprises the Interagency Task Force on Forests (ITFF)* and would be expanded to include other international forest-related organisations and convention secretariats. In addition to the mechanisms established by the United Nations, international non-governmental organisations such as the World Commission on Forests and Sustainable Development (WCFSD) provide perspectives and recommendations on sustainable management of forests.

Land Degradation and Agriculture Systems

A majority of the population in the Himalayan region depend on farming and agriculture as their primary source of sustenance. Hence efforts to address the problems of poverty, inequality, and marginalisation must begin with improving the well-being of mountain people by addressing the problems of mountain agriculture. Without improvements in the millions of small mountain farms, there will be little positive impact either on poverty or on the mountain environment. Since the cropping pattern as well as yield is largely determined by topographical and climatic conditions, the problems related to land and agriculture

---

* Established in 1995, the members of the ITFF include: the Centre for International Forestry Research (CIFOR), Food and Agriculture Organisation of the United Nations (FAO), the International Tropical Timber Organisation (ITTO), Secretariat of the Convention on Biological Diversity (CBD), the United nations Department for Social and Economic Affairs (UN/DESA), the United Nations Environment Programme (UNEP), the United Nations Development Programme (UNDP) and the World Bank.

in the Himalayan region vary with elevation, slope of the terrain, soil and climatic conditions. Land holdings here are much smaller as compared to those in the plains and cultivation entails the conversion of slopes into flatlands. Regular breaking down and alteration of the hill slopes have lead to severe land denudation and degradation problems. Land degradation in the Himalayas has also been caused by instabilities arising from the large-scale expansion of new roads and open cast mining (Stone, 1992).

Soil Erosion

Soil erosion, especially the loss of fertile topsoil either by gully, rill or sheet erosion, formation of hardpans and the exposure to indurated horizons has led to large-scale deterioration of the resource base of mountain agriculture. This is manifested in the decline in the per capita availability of crop-land, reduced availability of village commons (also called support land), fall in soil fertility and a decline in the carrying capacity of these areas. Large portions of arable land in the northern Himalayan states of Jammu & Kashmir and Himachal Pradesh have been rendered infertile and are in need of contingent soil and water conservation measures like contour farming, terracing, strip cropping, mulching, crop rotations, organic manuring, shallow ploughing etc. (Gupta and Banerjee, 1991). The rate of exploitation of land resources in these regions has been higher than the rate of replenishment.

The mountain people of the Himalayan region, especially those in the Northeast, over the years had adopted traditional practices of replenishment in the region. One such method was *Jhum* cultivation which is basically 'rotational bush fallow' agriculture. This traditional tribal practice enabled regeneration of forests before the same land was cultivated again. The *Jhum* cycle was once considered to be as long as 25 years, but in the recent past, studies have shown that the cycle has shrunk to as short as 4-5 years (Barthakur, 1981). As the *Jhum* cycle becomes successively shorter, the rate of soil erosion gets accelerated. This is a strong indicator of the deteriorating ecological balance of the region and is also a statement on the increasing human pressure on land and the growing food needs. Mountain agriculture in the Himalayas is entering a critical phase of transition as population demands for food and other agricultural commodities escalate rapidly and the damage to biological and physical systems supporting agriculture continues to accelerate.

High Value Crops

How farming of high-value crops (HVC) crops has increased food security and employment, thus improving the living conditions of mountain people has been shown by case studies, on agricultural transformation of some of the mountain areas. (Partap, 1995, Sharma, 1996, Sharma, S. 1997, Sharma and Sharma; 1997; Tulachan, 1997, and Badhani, 1998). They also show that accessibility, wider market network and strong R&D institutions are critical to the commercialisation of subsistence agriculture in the mountains through intensification of HVC crop cultivation. Studies also show that horticulture could be a profitable venture as the mountains have the potential for small-scale, specialised farming activities, which would yield high payoffs; the proper harnessing of such niche-based farming will help food security through direct use of products or trade in high value products (Jodha, 1991 and 1995). Such a move win have positive implications for the future development of mountain agriculture in terms of harnessing mountain comparative advantages—advantages which have positive ecological and economical implications.

Land and agriculture are the prime sources of livelihood in the Himalayan region and the problems of poverty, inequality and environmental degradation could be addressed by tackling problems associated with mountain agriculture. Efforts should be made to increase the productivity of the region. Balanced cultivation of food crops and commercial crops should be undertaken so that there is improvement in the value of agricultural output and food security. Emphasis to control soil erosion is very necessary.

Natural Disasters

Earthquakes

Earthquakes are a great threat to environmental stability and life in the Himalayan region as almost the entire region is prone to high seismic activity. The region has been hit by earthquakes of varying intensities in the past and similar threats remain imminent. The Himalayas were formed by a head on collision of the Indian and Eurasian plates, and the Indian plate continues to push the Asian plate northward at the rate of about 2 cm per year. This means that in every 100 years India moves 200 cm north against the Asian plate, and this colliding force builds up pressure continually for several years and is released in the form of earthquakes. The Indian Himalayas have experienced some significantly strong earthquakes in the last few decades. The major amongst them are the Kinnaur earthquake of 1975, Dharchula earthquake of 1980 and the Uttarkashi earthquake of 1991, which resulted in tremendous loss of life and property.

Landslides

Landslides are another major environmental hazard in the Himalayas, which have affected the entire mountain belt for many centuries. They normally occur in areas where the mountain slopes are extremely unstable, where vegetation is sparse and during the rainy season. Landslides are common both in the main Himalayas and in the barren cold desert regions. In early twentieth century overgrazing and deforestation were the main causes of landslides. Construction of roads have added its share of woes.

Avalanches

The term avalanche is defined as the failure of snow covered slopes, and takes place in areas of rapid accumulation of a huge mass of snow. During the process of accumulation, snow may fail or begin to slide downhill at a very fast pace due to an increase in stress which can be due to wind drifting, heavy snowfall in a short span of time, blasting, seismic activity or even thunder. The higher reaches of the Himalayas remain under a perpetual cover of snow and it is here that thousands of avalanches occur, involving the movement of thousands of tons of ice and vertical displacements of over 1500 m.

Solid Waste Generation and Management

Population settlement in mountain areas is concentrated in the valleys, which form a core area for all major commercial activities. This core area in most mountain towns has been provided with tapped water supply, piped sewage collection and discharge, and vats/dustbins for solid waste collection at the time of formation of these towns. However, as in other urban centers, population growth in hill areas has put pressure on urban resources and basic urban services which where primarily not designed for this level of population settlement. The management of solid waste in hilly towns has emerged as one of the most critical issues in term s of maintaining the environment in these towns. The infrastructure earlier created for solid waste management in most hilly towns fails to cater to the present level of solid waste generation and in some towns, about 50 per cent of the daily waste generation remains largely unattended (Darjeeling Municipality 2000).

Due to undulating mountainous terrain and narrow roads, it is difficult to employ large vehicles for garbage collection and to find large pieces of land to cater as disposal landfill. The initiatives for recycling and conversion of waste to energy, which would have otherwise reduced the quantum of waste to be disposed, are also not very evolved in hilly towns. The generation of solid wastes in Indian hill towns range from 5-6 tonnes per day for Mirik (West Bengal) to 20-25 tonnes for Darjeeling (West Bengal). The per capita waste generation in hilly towns ranges from 0.2 to 0.3 kg/person/day, which is lower than that in major Indian cities, which ranges from 0.2 to 0.6 kg/person/day.

Solid Waste Collection and Disposal

However, the disposal of solid wastes remains a critical problem in these towns due to less availability of flat land. In addition, tourist inflows in the hill towns generate a lot of packaging waste, which is difficult to dispose or degrade. Primary sources of solid wastes in Himalayan towns are residential areas, commercial market places, tourist places, hotels, restaurants, vegetable markets and slaughterhouses. Solid wastes are collected at community vats or disposal bins from where they are picked up either by sanitary trucks or tractor trolleys 3-5 times a day and disposed in disposal pits (Darjeeling Municipality, 2000).

Since the vats and dustbins are mostly provided on major roads in core areas of the towns, the waste generated from developmental areas (newly developing areas) of the town are not collected properly. They are mostly disposed in open drains, thus choking these drains or downstream water springs. The overall collection efficiency therefore remains around 50 per cent of the total solid waste generated. Population growth and urbanization are the most important factors contributing to growth in solid waste generation in the towns of the Himalayan region. In India the data on Himalayan towns like Darjeeling, Kurseong, Mirik and Kalimpong in the state of West Bengal show that the rate of population growth has been rapid in these towns since 1971. The population in these towns has been growing at average annual rates of around 2-5 per cent (Darjeeling Municipality, 2000).

Tourism is another major factor. Tourists flock to these towns but there is no arrangement for disposal of waste generated by them. Poor garbage collection in turn is causing serious health and environmental problems. Uncontrolled waste generation and accumulation on open spaces, wastelands and streets allow flies and vermin to spread diseases. The foul smell causes respiratory diseases, attracts pests (rats, mosquitoes, cockroaches, etc.) which may subsequently contaminate food. Indiscriminate disposal of waste also clogs drainage canals causing flooding during the rainy season. Leachate from decomposing and putrefying garbage also contaminates water sources (UNCHS, 1988). Improper handling and management of solid waste results in littering of these wastes, especially polythene bags and packing materials for foodstuff an over the landscape. This can negatively impact tourist inflows in the region.

Institutional Responses

Management of solid wastes in India is the prime responsibility of the municipal bodies and is governed by the Municipal Solid Wastes (Management and Handling) Rules, 2000. According to these rules, it is the responsibility of the respective municipal authority to implement the provisions of these rules and also undertake infrastructure development of related activities like collection, storage, transportation, treatment and disposal of municipal solid wastes.

Solid Waste Management Bodies

In addition, the 74th amendment of the Constitution of India identifies municipal bodies as agencies to provide services for effective solid waste management and also authorizes them to collect user fee for delivery of these services. Some initiatives have been introduced with the involvement of private sector for the processing of municipal solid wastes to useful products. Compost plants of capacities up to 500 tonnes per day have been set up in cities like Chennai and Delhi and may be established in hilly towns like Solan in Himachal Pradesh. However, in spite these initiatives, the present solid waste management system in the Indian Himalayan towns is largely ineffective, as a large amount of solid waste remains uncollected. There is some informal recycling activity present in these towns in the form of ragpickers who pick up papers, plastics and metal containers from the waste collection or disposal points. One such effort has been started by an entrepreneur, a high caste person in Norda, which engages about seventy rag pickers on a partnership basis.

In conclusion, it can be said that a solid waste management policy for hilly towns needs to be evolved which ensures focus on less use of waste generating material, discouraging use of plastic bags, maximizing resource recycling and reuse attempts to produce energy from the organic fraction of the waste. This would ensure that the pressure on existing landfill sites in these areas is reduced.

Mechanisms for Environmental Management in the Himalayan Region : Regional and International Responses

The previous sections have presented a discussion of the state of the environment in the Himalayan region, the key environmental problems, their causes, social and environmental impacts and some of the specific responses that have been undertaken to address these concerns. We now present an overview of the role of regional and international institutions and initiatives in the sustainable development of the Himalayan mountain ecosystem.

Regional Mechanisms

Arising from recognition of the environmental degradation of mountain habitats and the increasing impoverishment of mountain communities in the Hindu Kush-Himalayan region, the International Centre for Integrated Mountain Development (ICIMOD) was established in 1983 in Kathmandu. A unique regional institution with a focussed mandate of promoting sustainable mountain development, the Centre was set up based upon an agreement between His Majesty's Government of Nepal and UNESCO.

The Centre has eight members: Afghanistan, Bangladesh, Bhutan, India, Myanmar, Nepal and Pakistan. ICIMOD focuses its activities on environmental stability, sustainability of mountain ecosystems and poverty eradication in the Hindu Kush Himalayan region. In 1995 ICIMOD launched the First Regional Collaborative Programme for the Sustainable Development of the Hindu Kush Himalayas (1995-98). The programme was organized into:

- *Thematic groups:* Mountain Farming Systems, Mountain Natural Resources and Mountain Enterprises and Infrastructure
- *Service groups:* Mountain Environment and Natural Resources Information; Documentation, Information and Training; and Administration, Finance and Logistics

The Second Collaborative Programme for the Sustainable Development of the Hindu Kush Himalayas (1999-2002) has been launched with a focus on five major programme areas: promoting sustainable livelihoods for mountain households, gender balanced mountain development, sustainable management of mountain commons, capacity development of mountain development organizations and information and outreach. A regional response to the recommendations for sustainable development of mountain ecosystems proposed at the United Nations Conference on Environment and Development (UNCED) (please refer to the next section), came in the form of a Regional Conference on Sustainable Development of Fragile Mountain Areas of Asia. Held in 1994 in Kathmandu, Nepal, the conference brought together representatives from 20 Asian countries and international agencies.

SUDEMAA Call to Action

1. Poverty eradication and economic development by

- Encouraging the development of specialization of mountain areas in specific products and services
- Promoting sustainable mountain farming systems
- Improved access to extension and other services for rural areas
- Development of urban centers for marketing products and for providing centralized services
- Development of mechanisms to ensure that benefits derived from mountain resources accrue to mountain communities

2. Sustainable management of natural resources through

- the integration of indigenous knowledge in mountain development processes
- the enhancement of local capacities for mountain resource management, the restoration of ecologically-degraded lands
- the conservation and sustainable use and management of biodiversity

3. Gender-balanced decision-making in environment and development policies and programs through

- equality before the law
- the recognition of the economic value of women's work
- joint ownership of resources

Source: www.mtnforum.org/apmn/sudemma

SUDEMAA

The Conference adopted the Sustainable Development of the Mountain Areas of Asia (SUDEMAA) Call to Action. The recommendations of the SUDEMAA Call to Action are presented in BOX 2. The conference went on to recommend support mechanisms for the implementation of the SUDEMAA, these included:

- Establishment of National Mountain Task Forces/ Commissions/ Focal points

- Designation of focal points for mountain development by sub-regional, bilateral and multilateral agencies
- Exchange of information and regional collaboration among the mountain countries of Asia
- Capacity building
- Financing

As a follow up to the Regional Conference Sustainable Development in Fragile Mountain Areas in Asia, the Asia Pacific Mountain Forum (APMIV) was established in 1995 with its secretariat at the ICIMOD. An electronic forum and a regional node of the global Mountain Forum (please refer to the next section), the APMN has six sub-regional focal points in Australasia Pacific, North Central Asia, North East Asia, West Asia, South East Asia and the Hindu Kush Himalayas. These sub-regional focal points are located in academic and research institutions and promote networking in their sub-regions. Main activities of the APMN, through which it promotes exchange of information on issues highlighted by the SUDEMAA Call to Action, include publishing a biannual publication highlighting events and activities in the sub region, maintaining databases on regional mountain resources, facilitating study and exchange tours among mountain countries and organizing regional meetings.

International and Multilateral Initiatives

The United Nations Conference on Environment and Development (UNCED) held in 1992 at Rio recognized the crucial role played by mountain ecosystems by assigning a chapter ('Managing fragile ecosystems: Sustainable mountain development') to address concerns and evolve strategies for protecting and developing these ecosystems. Agenda 21, the road map for sustainability that emerged from the Rio conference highlighted that the livelihood of about 10 per cent of world population depended directly on mountain resources like water, forest and agriculture products and minerals. About 40 per cent of total world population lived in adjacent medium and lower watershed areas (Agenda 21 Chapter 13).

In addition, populations living in the valleys and plains depend on mountains for water as many major rivers originate there and climatic changes can alter hydrological regimes. The vulnerability of mountain areas to environmental degradation due to the fragile nature of the ecosystem, pressures placed by increasing population growth, tourism and economic development, was stressed in preparing a road map for sustainable development for these areas. Two programme areas for sustainable mountain development were identified in 1992:

- Generating and strengthening knowledge about the ecology and sustainable development of mountain ecosystems
- Promoting integrated watershed development and alternative livelihood opportunities

Joint Effort

The institutional mechanism envisaged to implement these programs primarily stressed action by national governments, In coordination with relevant international and regional organizations, on the following fronts:

- Strengthening existing institutions or establishing new institutions at local, national and regional levels. For example, by establishing task forces or watershed development committees that complement existing institutions and support local community initiatives.

- Promoting local participation through the adoption of national policies and legislation that provide local communities with an incentive to undertake conservation measures

- Creating mechanisms for cooperation and information exchange among national and regional level institutions for building and maintaining databases and information systems on ecological resources, evaluation of environmental risks and natural disasters in mountain areas. Promoting regional cooperation and exchange of information was particularly stressed for countries sharing mountain ranges and watersheds and those vulnerable to natural disasters.

- Regional and international cooperation and exchange of information between international institutions such as the World Bank, the IFAD (International Fund for Agricultural Development), regional institutions such as the ICIMOD and other non-governmental and research institutions working in the area of mountain development,

- Promoting the co-ordination of regional efforts to protect fragile mountain ecosystems, through the creation of appropriate mechanisms and instruments, including regional legal instruments.

Sustainable Mountain Development

In the ten years since the Rio Summit, substantial progress has been made in establishing international and regional networks of institutions dedicated to the issue of sustainable mountain development. An overview of the mandate of two of these institutions, the Inter-Agency Group on Mountains and the Mountain Forum is presented below :

Established after the UNCED, the Inter-Agency Group on Mountains is the main institutional mechanism for advising on the implementation of activities proposed in Chapter 13 of Agenda 21. It comprises participants from both the United Nations and other organizations, with the Food and Agriculture Organization (FAO) as the coordinator.*

The Mountain Forum was established in 1995 through international collaboration from non-government organizations, universities, governments, multilateral agencies and the private sector. The forum has enabled generate increased awareness and exchange of information on mountain issues through email discussion lists, electronic conferencing and online library resources. It comprises five regional networks, with membership from over 100 countries.

Conclusion

Environmental sustainability of the Himalayan region is critical to the well being of millions of people in the Indian subcontinent who are directly or indirectly dependent on its resources. Increased accessibility, population pressures, economic development and the concomitant excessive utilisation of the regions resources have resulted in a threat

* The FAO is the Task Manager for Chapter 13 of Agenda 21.

to the health of this mountain ecosystem and its inhabitants. Local, national and international efforts for environmental protection of the region have been made, which is highlighted by the popular environmental movements, policy and legislation, research networks and institutions that have sought to address different environmental and social issues in the region. However, despite these efforts, the Himalayan region continues to face increasing environmental degradation.

Better environmental management will have to be based on an understanding of the specific social and economic context, environmental specificity's and the critical geopolitical role of this region.

Some recommendations for improving the state of the environment in the Himalayan region are:

- Creating mechanisms for maximising community participation in the design, implementation and monitoring of environmental protection programmes.
- Establishing mechanisms for co-ordinating between local, national and regional institutions in the region.
- Developing stronger national, regional and international co-operation to address the needs of mountain communities.
- Increasing international development assistance for the region.

The year 2002 has been declared as the International Year of the Mountains by the United Nations General Assembly. This is thus an opportune time to recognise the crucial role of the Himalayan mountain ecosystem in the region and the world, strengthen existing initiatives and develop new mechanisms for sustainable mountain development, both at the national and international level.

References

Badhani, KN, 1998. Enterprise-based transformation of hill agriculture: a case study of vegetable growing farmers in Garampani area, Nainital District, India. Discussion Paper, Series No. MEI 98/5, Kathmandu, Nepal, ICIMOD.

Bahadur, J, 1998a. Himalayan eco-hydrology-an emerging topic. Journal. Indian Association of Hydrologists, Roorkee.

Bahadur, J, 1998b. Himalayan Glaciers. www.mtnforum.org/resources/library.

Bahadur, J, 2000. Water resources management in the Himalayan region. Proceedings of the Internal Conference on Managing natural resources for sustainable agricultural production in the 21st Century, New Delhi, Feb. 14-18, 2000, PI). 459-461

Barthakur, LK, 1981. Shifting cultivation and economic change in the north-eastern Himalaya. The Himalaya: aspects of change. Ed. J S Lall, Oxford University Press, Delhi. pp. 447-460.

Biringer, KI, 1998. Siachen Science Centre: a concept for cooperation at the top of the world. Report, Sandia National Laboratories, U S Department of Energy. www.cmc.sandia.gov.

Bourque, Susan and Kay Barbara Warren, 1980 Women of the Andes: patriarchy social change in two Peruvian towns. Ann Arbor, University of Michigan Press.

Center for Women and Development (CWD), 1988. Women's work and family strategies under conditions of agricultural modernization. CWD, Kathmandu, Nepal.

Darjeeling Municipality, 2000. Integrated Development Scheme. Darjeeling Municipality.

Dhar, U, 1996. Overview of Himalayan biodiversity. Changing perspective of biodiversity status in the Himalaya, New Delhi. The British Council.

Farooquee, Nehal A. and Rawat, D.S. 1997. Development Dilemma, Indian Scenario and Rural Himalaya (A Central Himalayan Perspective). G.B. Pant Institute of Environment and Development. Almora, India. Himavikas Publication No. 11, p. 128.

Fletcher, Susan 1995. International Forest Agreements: Current Status. Congressional Research Service Report for Congress. National Council for Science and the Environment. Washington, D.C. Accessed from http:/ / www.cnie.org/nle.for-11.html

FSI 2000. State of Forest Report 1999. Forest Survey of India (Ministry of Environment and Forests), Government of India, Dehradun, p. 113.

Fujita, Koji, Masayoshi Nakawo, Yoshiyuki Fuji and Prem Paudyal, 1997. Changes in glaciers in Hidden valley, Mukut Himal, Nepal Himalayas, from 1974 to 1994. Journal of Glaciology, Vol. 43, No. pp. 145, 583-588.

Gergan, J T, Dobhal, D P and Kaushik Rambir, 1999. Ground penetrating radar ice thickness measurements of Dokriani bamak (glacier), Garhwal Himalayas. Current Science, Vol. 77, No. 1, pp. 169-173.

Ghosh, A.K., 1997. Himalayan Fauna with special reference to endangered and endemic species. Himalayan Biodiversity: Action plan. GB Pant Institute of Himalayan Environment and Development.

Gupta, R. K., 1983. In the living Himalayas: aspects of environment, resource and ecology of Garhwal. Today and Tomorrow Printers and Publishers.

Gupta, R. D. and Banerjee, S. Y., 1991. Problems and management of soil and forest resources of north-west

Himalayas. Mahajan Book Center and Publishers, Jammu. p. 162.

Hasnain, S. I., 1999, Runoff characteristics of a glacierized catchment, Garhwal Himalaya, India Hydrological Science Journal, Vol. 44(6), pp. 847-854.

Human, Joe and Pattanaik, Manoj, 2000. Community forest management: a Casebook from India. Oxfam Great Britain. p. 176.

IPCC, 2001.

Climate Change 2001: Impacts, Adaptation and Vulnerability. Contribution of Working Group 11 to the Third Assessment Report of the Intergovernmental Panel on Climate Change. James J McCarthy, Osvaldo F Canziani, Neil A Leary, David J Dokken, Kasey S White; editors. Cambridge University Press. United Kingdom. pp. 533-590.

Ives, J.D and Messerli, B, 1989. The Himalayan dilemma: reconciling development and conservation. Routledge, London, UK, p. 366.

Jain, S. 1991. Standing up for the trees: women's role in Chipko movement. Sally Sontheimer (Ed.). Women and the environment: a reader. New York: Monthly Review Press.

Jodha, N.S., 1991. Sustainable agriculture infragile resource zones: technological imperatives. Economic and Political Weekly, March 30, 1991.

Jodha, N.S., 1995. Enhancing food security in a warmer and more crowded world: factors and processes infragile zones. Climate Change and World Food Security. U.K. Downing Thomas E. Springer, NATO Scientific Affairs' Division.

Jodha, N. S. 2001. The Nepal middle mountains. www.unu.edu/unupress/unubooks/uul4re/uul4reog.htm. Downloaded 8 August 2001.

Jodha, N.S, 1991. 'Sustainable agriculture in fragile resource zones: technological imperatives'. Economic and Political Weekly. March 30, 1991.

Jodha, N.S, 1995. Enhancing food security in a warmer and more crowded world: factors and processes infragile zones. Climate change and world food security. U.K, Downing Thomas E. Springer, NATO Scientific Affairs Division

Kawosa, M.A, 1988. In Remote Sensing of Himalayas. Natraj Publications, Dehradun.

Khoshoo, T. N. 1992. Govind Ballabh Pant Memorial Lecture: II-14th.

September 1992. G B Pant Institute of Environment and Development. Almora, India, p. 129.

Mehta, 1994. The Transformation of subsistence agriculture and gender inequalities in a Central Himalayan valley, Uttar Pradesh, India. Unpublished dissertation, Boston University, UMI, Ann Arbor.

Mehta, Manjari, 2001. Cultural diversity in the mountains: issues of integration and marginality in sustainable development. www.mtnfurum.org/resources/library/mehta95a.htm and www.mtnfurum.org/resources/library/mehtag5b.htm, downloaded on 28 August 2001.

MoEF, 1999. National Forestry Action Programme-India. Volume 1: Status of Forestry in India. Ministry of Environment and Forests, Government of India. New Delhi, pp. 177.

Murthy, V K, 1978. Environmental problems of water resource development in the Himalayan region. Proceedings of National Seminar on Resource development in the Himalayan region, New Delhi, April 10-13,1978. Dept. of S&T, Govt. of India, pp. 58-69.

Mool, Pradeep, 2001. Mountain flash Moods: Glacial lakes and glacial lake out burst flood events in the Hindu Kush Himalayan region. ICIMOD, Newsletter No. 38.

Negi, S.S., 1991. Himalayan rivers, lakes and glaciers. Indus Publishing Co. New Delhi, p. 182.

Oerlemans, J, 1994. Quantifying global warming from the retreat of glaciers. Science, 264, pp. 243-45.

Partap, T., 1995. High Value Cash Crops in Mountain Farming: Mountain Development Processes and Opportunities. Kathmandu, Nepal, ICIMOD.

Rizvi, M, 2001. Forecasting water flows in Pakistan's Indus river. www.idrc.ca/reports

Rodgers, W.A and Pawar, H.S, 1988. Planning wildlife protected area network in India, Vol. 1&2. WII Dehradun.

Sands, Phillipe, 1996. Bangladesh-India: treaty on sharing of the Ganges water at Farakka. www.asil.org/ilm/india.htm

Schuler Sydney, 1981. The women of Baragacn. The status of women in Nepal: indigenous agriculture in rural Nepal. Mountain Research and Development. 5(l). pp. 31-44.

Sharma, K. P., 1993. Role of rainwater in major river systems of Nepal. International symposium on snow and glacier hydrology [Young, G.J. (ed.)]. International Association of Scientific Hydrology, Kathmandu, Nepal, pp. 113-22.

Sharma, H.R., 1996. Mountain Agricultural Development Processes and Sustainability-Micro-level Evidence from Himachal Pradesh, Indian Himalayas. Kathmandu, Nepal, ICIMOD.

Sharma, K.P., 1997. Impact of Land Use and Climatic Changes on Hydrology of the Himalayan Basin: a case study of the Kosi Basin. Diss. University of New Hampshire, Durham, NH, U.S.A. p. 247.

Sharma, S., 1997. Agricultural Transformation Processes in the Mountains of Nepal: empirical evidence from Ilam District. Kathmandu, Nepal, ICIMOD.

Sharma, H.R., and Sharma E., 1997. Mountain agricultural transformation processes and sustainability in the Sikkim Himalayas. India. Kathmandu, Nepal. ICIMOD.

Singh 1992. Man and forest interactions in Central Himalaya. Himalayan environment and development: problems and prospects. G B Pant Institute of Environment and Development. Almora, India. Himavikas Occasional Publication No. 1. pp. 57-79.

Singh, D.K & Hajra, P.K. 1996. Floristic diversity: changing perspective of biodiversity status in the Himalaya, New Delhi. The British Council.

Singh P, S K Jain, and N Kumar, 1997. Estimation of Snow and glacier melt contribution to the Chenab River, Western Himalayas. Mountain Research and Development, 17(1), 49-56.

Singh, A K and Hasnain, S 1, 1998. Major ion chemistry and weathering control in a high altitude basin: Alaknanda river, Garhwal Himalaya, India. Journal of Hydrological Sciences, 43(6), pp. 825-843.

Stone, Peter B, 1992. State of the World's Mountains-A Global Report. Mountain Agenda. Peter B Stone, editor, p. 391.

Swaminathan, M. S. 1992. The First Pandit Govind Ballabh Pant Memorial Lecture. Himalayan environment and development: problems and perspectives. G. B. Pant Institute of Himalayan Environment and Development. Gyanodaya Prakashan, Nainital. pp. 13-22.

TERI 1992. Energy, environment and sustainable development in the Himalayas. Proceedings of the National Workshop on Energy and environmental issues in Mountain development. Held on 24-25 September 1991 in Shimla, India. Pradeep Monga and P Venkata Ramana, editors. New Delhi, p. 202.

Tulachan, P.M., 1997. Ensuring local food security: an example from the mountains of Nepal. Issues in Mountain Development 97/1. Kathmandu, Nepal, ICIMOD.

UNCHS. Refuse collection of the vehicle for the developing countries, HS/138/88E, UNCHS (Habitat) Nairobi, Kenya, 1998.

United Nations 2001. About the intergovernmental process. Accessed from http://www.un.org/esa/sustdev/aboutiff.htm

Upadhaya, D S, 1995. Cold climate hydrometeorology, New Age International (P) Ltd. Publishers, New Delhi. p. 345.

Verghese B G and Ramaswamy R Iyer, 1993. Harnessing the eastern Himalayan rivers, regional cooperation in Southeast Asia. Center for Policy Research, Konark Publishers Pvt. Ltd., New Delhi. p. 180.

Vohra, C.P, 1981. The climate of the Himalayas. The Himalaya: aspects of change [Lall, J.S. (ed.)] Oxford University Press, New Delhi, India, pp. 138-151.

Vohra, C.P, 1993. Himalayan Glaciers. Harnessing the eastern Himalayan rivers. Regional Cooperation in South Asia. Ed. B G Verghese & Ramaswamy R Iyer, Kenark, New Delhi.

2

History of Himalayan Ecology

Civilisational Identity

The Himalayas are not merely a geographical feature, a range of mountains; they epitomise a people's civilisational identity that goes back to the dawn of history. If these majestic mountains were not there, the rain clouds sweeping up from the Indian Ocean would have passed over the Indian subcontinent into central Asia leaving it a burning desert.

Archaeological evidence reveals that urban civilisation first emerged in the valleys of the rivers coursing down from the Himalayan mountains. When man elsewhere was cautiously emerging from his cave shelters, spectacular cities flourished at Mohenjodaro and Harappa in the Indus valley.

Climate of Himalayas

Himalayan Coverage Area:	2,250 km
Average Width:	200 km
Himalayan Forests:	Pine, Deodar, Fir, Oak, Rho-dodendron, Birch
Monsoon Season:	Mid-June till the end of September

Saga of Mountains, Forests, Pastures

The Himalayas with dazzling pinnacles of snow-covered ranges extend for 2,250 km from the Namcha Barwa on the bend of the Tsang-po (Brahmaputra) to Nanga Parbat on the Indus. The range runs east to west up to central-Nepal and then takes a southeast to northwest direction.

The average width of the Himalayas is about 200 km. The compositions of the forests on this mighty range are effected by the combined effect of altitude, rainfall and latitude. The rainfall, mainly from the monsoons, decreases from the east to the west. Comparatively, in winter, due to the influence of tropical west wind drift, the northwest areas have more precipitation in the form of rain and snow.

Climatic Variation in the Himalayas

From these snow covered ranges with big glaciers, flow the great rivers that have made the Indo-Gangetic plain one of the most fertile in the world and the heart Land of India. The Himalayan Alpine climate varies according to the elevation. It gets colder as the elevation increases and gets wetter as the elevation drops. As a result the temperature and climatic changes in the Himalayan regions change very quickly.

All of a sudden there can be occurrences of monsoons, floods, high winds, snowstorms and other types of precipitation, which makes the climate over here quiet an unpredictable and dangerous one.

The two major seasons of the Himalayan region are winter and summer. During the winter the region receives the maximum snow with very icy temperatures. Summer are quiet mild over here, making the places over here quiet good summer holiday hideouts. Usually throughout the year the Alpine Himalayan region receives snow.

Beyond the Monsoon

The regions such as Ladakh and Zanskar that lie to the North of the main Himalayan range and escape the full impact of the monsoon. Humidity is always low in these regions, and receives only a few centimeters of rainfall a year. These regions also experience some of the coldest temperatures anywhere in the world, and it doesn't get any warmer till the spring season in late April or early May. In June, daytime temperatures frequently rise to the mid 200Cs, the snow on the passes melts and most of the treks can be undertaken from then on until the middle of October.

Heavy rainstorms can occasionally be experienced in July and August, and River crossing should be undertaken with great care at this time. By September the conditions are ideal, and they normally remain so until late October even though nigh-time temperatures may fall below freezing. By November, the early winter snows fall on the passes closest to the Himalayas. In winter the villagers still travel, enduring the intense cold, to follow the valley floors where River crossings are no longer a problem.

Change in Weather

After October the daytime temperatures drop, but the weather is generally settled until the middle of November when the first of the heavy winter snows fall on the high mountain passes.

Winter months from December to March are often bleak. April and May are characterised by heavy precipitation, which falls as snow in the mountains, precluding trekking over the passes until the spring snows melt in June. Most of the hill states of the West Himalayas including Kangra and Chamba, the Kullu valley, Shimla in Himachal and most regions of the Garhwal and Kumaon in Uttaranchal come under the influence of the Indian monsoon. Both Darjeeling and Sikkim are subject to the Indian monsoon that sweeps up from the Bay of Bengal, bringing heavy rainfall from early June until the end of September. The post-monsoon months of October and November provide settled conditions, with clear views of the mountains, although nighttime temperatures above 3,500 m frequently fall below freezing.

Forest Cover-The Glory of the Himalayas

Because of the constant most of the Himalayan regions have a harsh environment therefore few animals and plants can survive over here.

The few plants that do inhabit the Alpine consist of rhododendrons, the tea plant and shrub type plants. They have to adapt to the freezing temperatures, high winds and to a short growing season. That is why most of the plants grow low to the ground. The Hindustan-Tibet road now called National Highway No. 22 takes one right across the Himalayas from the plains over forested mountain ranges, along the Sutlej through the gorge in the great Himalayan range near Shipkila on the Indo-Tibet border.

One of the most interesting and accessible areas are the forests on the ridge line that form the Indo Gangetic watershed on which lie Shimla, Narkanda and the most beautiful forests of silver Fir starting from Narkanda. Part of the Hindustan-Tibet road run along this watershed ridge. From the plains to the snow clothed mountains, one goes through fine forests of Pine, Deodar, Fir, Oak, Rhododendron, Birch and finally between the tree line and the snows, rich alpine pastures.

Inhabited areas, on cultivated terraced fields, can be seen women in their customary colourful clothes decorating the hillside like alpine blossoms. The mountain fauna that are found in the Himalayan Alpine are similar to the mountain animals found in the surrounding areas of the Himalayas. Some animals that have adapted the Himalayan climate include the mountain goat, which has a thick coat for warmth and strong hooves for running up the rocky slopes.

The Awesome Himalayan Panorama

To know the mountains one has to be amongst them. In the evening as one begin settling down so do the mountains, valleys and clouds. It is a wonderful sight to see a sea of clouds sinks in repose in the-deep valleys. At the same time the setting sun converts the surrounding peaks into molten gold which gradually fade away into pale pink and steel grey. At night, the moon lends a peculiar charm to the snow transferring the rugged peaks and ridges into soft contours. The hamlets in the valleys, surrounded by fields, are in deep slumber. The silence of the mountains is complete. The only sound one can hear is of the campfire.

In the morning, the rising glow of the sun reflects from peak to peak and then descends to the lower slopes awakening all. The sun's warmth rouses the clouds to gradually climb the mountains, hugging the valley sides as they rise and engulf the peaks with their crowns of snow. This is an experience so vivid and revitalizing that it can never be forgotten.

Natural Warning

In recent years, there has been great emphasis on saving the trees but the actual progress has been insignificant. Day by day, the trees are being destroyed and being cut down without replacement and without a thought for our children's future. Overgrazing and ruthless assault on herbs is destroying the meadows and pastures.

Absence of Forests

In the absence of forests, there will be less precipitation in the form of rain and snow-leaving the grand peaks without their white mantle. This ecological damage, if not redeemed, will lead to catastrophes in the form of extreme floods, droughts, forests fires and disappearance of wildlife.

The diversity of the plants, herbs, flowers, landscapes is truly stunning and reflects Nature's brilliant ecological plan. The lines taken from Devistotra, a Hindu Shastra, (500 BC) presents quite a dear picture of the Himalayan lands: "So long as this land will have mountains, forests and pastures that long the earth will survive sustaining you and the coming generations."

Himalayan Mountain Range

Range: Main Himalayan Range, Pir Panjal Range, Dhaula Dhar Range, Siwalik Hills, Zanskar Range, Ladakh Range, & East Korakoram. The Himalayas is one of the youngest mountain ranges in the world. Its revolution can be traced to the Jurassic Era (80 million years ago) when the world's landmasses were split into two: Laurasia in the Northern hemisphere, and Gondwanaland in the southern hemisphere. The landmass which is now India broke away from Gondwanaland and floated across the earth's surface until it collided with Asia. The hard volcanic rocks of India were thrust against the soft sedimentary crust of Asia, creating the highest mountain range in the world.

The River System

It was a collision that formed mountain ranges right across Asia, including the karakoram, the pamirs, the Hindukush, the Tien Shan and the Kun Lun. The Himalayan mountains, at the front of this continental collision, are still being formed, rising and assuming complex profiles. For the ancient geographer, the complexities of this vast mountain range were a constant source of speculation.

From the earliest accounts, Mt. Kailash was believed to be the centre of the universe with the River systems of the Indus, the Brahmaputra, and the Sutlej all flowing from its snowy ridges and maintaining the courses, which they had followed prior to the forming of the Himalaya.

The Sutlej was able to maintain its course flowing directly from Tibet through the main Himalaya range to the Indian subcontinent, while the huge gorges on both flanks of the Himalaya reflect the ability of the Indus and the Brahmaputra to follow their original courses.

The Indus

The Indus flows west until it rounds the Himalaya by the Nanga Parbat Massif, while the Brahmaputra flows eastwards for nearly 1000 kms around the Assam Himalayas and descends to the Bay of Bengal. It was not surprising, therefore, that 19th century geographers experienced formidable difficulties n tracing the River systems, and defining the various mountain ranges that constitute the Himalaya. Even today, with the advent of satellite pictures and state-of-the-art ordnance maps, it is still difficult to appreciate the form and extent of some of the ranges that constitute the Himalaya.

Main Himalaya Range

This is the principal mountain range dividing the Indian subcontinent from Nanga Parbat in the west, the range stretches for over 2,000 km to the mountains bordering Sikkim and Bhutan in the east. The west Himalaya is the part of this range that divides Kashmir and Himachal Pradesh from Ladakh. The highest mountains here are Nun and Kun. In Kashmir the subsidiary ridges of the Himalaya include the North Sonarmarg, Kolahoi and Amarnath ranges. Further east, the Himalaya extends across to the Baralacha range in Himachal Pradesh before merging with the Parbati range to the east of the Kullu valley. It then extends across kinnaur Kailas to the swargarohini and Bandarpunch ranges in Uttaranchal.

Further east it is defined by the snow capped range North of the Gangotri glacier and by the huge peaks in the vicinity of Nanda Devi, the highest mountain in the Indian Himalaya. In Western Nepal the range is equally prominent across the Annapurna and Dhaulagiri massifs, while in Eastern Nepal the main ridgeline frequently coincides with the political boundary between Nepal and Tibet.

Major Passes

The major passes over the main Himalaya range include the Zoji la, at the head of the Sindh valley; the Boktol pass, at the head of the Warvan valley; the Umasi la in the Kishtwar region; and Thekang la and the Shingo la between Lahaul and the Zanskar region of Ladakh.

It also includes the Pin Parbati pass between Lahaul and the Zanskar region of Ladakh. It also includes the Pin Parbati pass between the Kullu valley and Spiti, while in Kannaur it is transversed when crossing the charang la in the Kinnaur Kailash range. In Uttaranchal, roads are being constructed to the main places of pilgrimage in the heart of the Himalaya. These include Yamunotri and the source of the Yamuna River, Gangotri at the head of the Bhagirathi valley, Kedarnath at the head of the Mandakini valley, and Badrinath in the Alaknanda valley.

There are, however, many trekking possibilities across the mountain ridges and glacial valleys including those bordering the Nanda Devi sanctuary. The main Himalaya range extends east across central Sikkim from the huge Kangchenjunga massif, which includes Kangchenjunga 1, the world's third highest peak. The east Himalaya is breached by the headwaters of the Tista River, which forms the geographical divide between the verdant alpine valleys to the south and the more and regions that extend North to Tibet. Trekking possibilities are at present confined to the vicinity of the Singali ridge, an impressive range that extends south from the main Himalaya and forms the border between India and Nepal. In Darjeeling the treks include the route along the southern extremity of the Singali range, while in Sikkim the trails out of Yuksom explore the ridges and valleys to the south to the Kangchenjunga massif.

Pir Panjal Range

The Pir Panjal Range lies south of the main Himalaya at an average elevation of 5,000m. From Gulmarg in the North west it follows the southern rim of the Kashmir valley to the Banihal pass. Here the Pir Panjal meets the ridgeline separating the Kashmir valley from the Warvan valley. From Banihal the Pir Panjal sweeps south-east to Kishtwar, where the combined waters of the Warvan and Chandra Rivers meet to form the Chenab River, one of the main tributaries of the Indus.

Passes in Pir Panjal

The main passes over the Pir Panjal include the pir panjal pass due west of Srinagar, the Banihal pass which lies at the head of the Jhelum River at the southern end of the Kashmir valley, and the sythen pass linking Kashmir with Kishtwar. In Himachal Pradesh the main passes are the Sach which links the Ravi and the Chandra valleys, and the Rohtang, which links the Beas and Kullu valleys with the upper Chandra valley and Lahaul. Roads are constructed over all these passes.

The Banihal is now tunnelled and another road has been made over the Sythen pass in Kashmir and the Sach pass in Himachal Pradesh. For trekkers there is still the attraction of the Kugti, Kalicho, and Chobia passes between the Ravi valley and Lahaul, and the Hampta pass links the Kullu valley with Lahaul.

Dhaula Dhar Range

The Dhaula Dhar range lies to the south of the Pir Panjal. It is easily recognised as the snow-capped ridge behind Dharamsala where it forms the divide between the Ravi and the Beas valleys. To the west it provides the divide between the Chenab valley below Kishtwar and the Tawi valley which twists south to Jammu. This is the range crossed at Patnitop on the Jammu-Srinagar highway. To the east it extends across Himachal Pradesh forming the high ridges of the Large gorge and extending south of the Pin Parvati valley before forming the impressive ridgeline east of the Sutlej River.

Thereon it forms the snow capped divide between the Sangla valley and upper tons catchment area in Uttaranchal, including the Har Ki Dun Valley. Beyond the Bhagirathi River it forms the range between Gangotri and Kedarnath before merging with the main Himalaya at the head of the Gangotri glacier. There are many attractive trekking passes over the Dhaula Dhar. These include the Indrahar Pass North of Dharamsala: and in Kinnaur, the Borasu pass linking the Sangla valley to Har-ki-Dun in Uttaranchal.

Siwalik Hills

The Siwalik Hills, also known as Shiwalik Hills, lie to the south of the Dhaula Dhar, with an average elevation of 1,500 to 2,000 m. They are the first range of hills encountered en route from the plains and are geologically separate from the Himalaya. They include the Jammu hills and Vaishno Devi, and extend to Kangra and further east to the range south of Mandi. In Uttaranchal, they extend from Dehra Dun to Almora before heading across the southern borders of Nepal. Most of the range is crossed by a network of roads, linking the Northern Indian plains with Kangra, the Kullu valley, Shimla and Dehradun.

Zanskar Range

The Zanskar range lies to the North of the main Himalaya. It forms the backbone of Ladakh south of the Indus River, stretching from the ridges beyond Lamayuru in the west across the Zanskar region, where it is divided from the main Himalaya by the Stod and Tsarap valleys, the populated districts of the Zanskar valley. The Zanskar range is breached where the Zanskar River flows North, creating awesome gorges until it reaches the Indus River just below Leh. To the east of the Zanskar region the range continues through Lahaul and Spiti, providing a complex buffer zone between the main Himalaya and the Tibetan plateau.

It continues across the North of Kinnaur before extending west across Uttaranchal, where it again forms the intermediary range between the Himalaya and the Tibetan plateau, which includes Kamet, the second highest peak in India. The range finally peters out North east of the Kali River-close to the border between India and Nepal. On the Zanskar range, the Fatu La, on the Leh-Srinagar road, is considered the most easterly pass; while the Singge La, the Cha Cha La and the Rubrang La are the main trekking passes into the Zanskar valley. For the hardy Ladakh trader, the main route in winter between the Zanskar valley and Leh is down the icebound Zanskar River gorges. Further to the east, many of the Zanskar range passes to the North of Spiti and Kinnaur are close to the India-Tibet border, and are closed to trekkers.

Ladakh Range

The ladakh range lies to the North of Leh and is an integral part of the Trans-Himalayan range that merges with the Kailash range in Tibet. The passes include the famous Kardung La, the highest motorable pass in the world, while the Digar La to the North east of Leh is at present the only pass open to trekkers.

East Karakoram Range

The East Karakoram Range is the huge range that forms the geographical divide between India and Central Asia. It includes many high peaks including—Teram Kargri, Saltoro Kangri and Rimo, while the Karakoram Pass was the main trading link between the markets of Leh, Yarkand and Kashgar. At present this region is closed to trekkers, although a few foreign mountaineering groups were permitted to climb there in the last decade.

3

The People and Tribes in Rural and Urban Ecosystems

Negroids, Mongoloids and Aryans

The Inhabitants and the Migrants

The Hindu epics and Puranas refer to the original inhabitants of the Himalayas — the Kulinds, Kiratas and Kilinds and Kinnars. Later Khasas and the Darads are also mentioned. However, Kashmir and other parts of Himalayas right upto Northeast India have been a happy ground for migrations from Central Asia and Tibet. Today three ethnic types constitute the Himalayan population: Negroids, Mongoloids and Aryans.

From very early times there have been migrations into the Himalayas and within it. Spiritual quest motivated a few to migrate there, and a small minority responded to the call of these mountains to test their own endurance and will power. Pursuit of profit propelled others. Reasons of state dictated the posting of garrisons even in remote, desolate areas. All these factors combined over a period of time to change significantly the complexion of the local populace, There have been waves of migration from Nepal to Sikkim and Kumaon, for instance, and from Tibet. At present it is extremely difficult to separate the different racial strains.

Bhutiyas of Bhutan

The Bhutanese are Bhutiyas of Mongolian origin who refer to themselves as Drukpa-inhabitants of Druk Yul or the 'Land of the Thunder Dragon'. Apart from a few obscure areas of Nepal and Ladakh, and Spiti in India, the Bhutanese are the only large group to follow traditional Buddhism. It has been a closed land for tourism for more than a century. Now that the roads are being constructed by the government, every care is taken that the cultural fabric is not disturbed.

People of Sikkim

The Sikkimese consist of three different groups-the Lepchas, the Bhutiyas and the Nepalis. The Lepchas are the original inhabitants but are now in a minority. Not much is known of their history before their conversion to Buddhism and the enthronement of Phuntsok Namgyal as the first historic ruler of Sikkim.

Nepali People

The people of Nepal are a complex mix of racial patterns. The dominant Hindu castes of Brahmin, Thakur and Chetri, along with several others speak Nepali. They are the descendents of Raputs who migrated from the north Indian states during the Sultanate rule. The Gurungs, Magars, Tamaings, Rais and Limbus form the Gurkha regiments of the British and Indian armies. These are part of the mongoloid, tribally organized groups of hill farmers who dominate the middle hills. The Sherpas of the Solo Khumbu region in the northeast of the country are among the many Bhubya groups who speak dialects of Tibetan.

People of Himachal Pradesh

In Kumaon and Garhwal in the central Himalayas, Khasas and Dorris were the original inhabitants. The Khasas, historians surmise, were a west Central Asian nomadic tribe who entered through the northwest and spread from Kashmir to Assam.

People of Uttaranchal

In Himachal Pradesh, the descendants of these Khasas are known as "Kanets" and now claim Rajput status. The majority of the population in the present-day central Himalayas has Khasa ancestry. Immigrant Brahmins and Kshatriyas from the plains brought caste division with them and introduced new forms of social organization rooted in Hindu orthodoxy.

The Ladakhis

The Ladakhis are of ethnic stock different from that of the people of Kumaon and Garhwal. According to folklore, Ladakh was once totally populated by Darads. The latest archaeological finds give credence to this popular belief. The Mons belonging to the Mongoloid stock, who are now far more numerous there, seem to have migrated at a much later date.

Most people in the Himalayas sustain themselves by a combination of agriculture and animal husbandry. Until very recently, those inhabiting the higher reaches migrated to lower altitudes during the winter months.

Indian Tribes

Composition and Location-India tribal Population

The original inhabitants of India constitute roughly 8 per cent of the nation's total population, nearly 68 million people according to the 1991 census. The other 92 per cent whose predecessors have been, for a long time migrating from Central Asia, Tibet and the East and Southeast. One concentration lives in a belt along the Himalayas stretching through Jammu and Kashmir, Himachal Pradesh, and Uttar Pradesh in the west, to Assam, Meghalaya, Tripura, Arunachal Pradesh, Mizoram, Manipur, and Nagaland in the northeast. Another concentration lives in the hilly areas of central India (Madhya Pradesh, Orissa, and, to a lesser extent, Andhra Pradesh); in this belt, which is bounded by the Narmada River to the north and the Godavari River to the southeast tribal peoples occupy the slopes of the region's mountains.

Other tribals, the Santals, live in Bihar and West Bengal. There are smaller numbers of tribal people in Karnataka, Tamil Nadu, and Kerala, in western India in Gujarat and Rajasthan, and in the union territories of Lakshadweep and the Andaman and Nicobar Islands.

The extent to which a state's population is tribal varies considerably. In the northeastern states of Arunachal Pradesh, Meghalaya, Mizoram, and Nagaland, upward of 90 per cent of the population is tribal. However, in the remaining northeast states of Assam, Manipur, Sikkim, and Tripura, tribal peoples form between 20 and 30 per cent of the population. The largest tribes are found in central India, although the tribal population there accounts for only around 10 per cent of the region's total population. Major concentrations of tribal people live in Maharashtra, Orissa, and West Bengal.

The Tribals

In the south, about 1 per cent of the populations of Kerala and Tamil Nadu are tribal, whereas about 6 per cent in Andhra Pradesh and Karnataka are members of tribes. There are some 573 communities recognized by the government as Scheduled Tribes and therefore eligible to receive special benefits and to compete for reserved seats in legislatures and schools. They range in size from the Gonds (roughly 7.4 million) and the Santals (approximately 4.2 million) to only eighteen Chaimals in the Andaman Islands. Central Indian states have the country's largest tribes, and, taken as a whole, roughly 75 per cent of the total tribal population live there.

Apart from the use of strictly legal criteria, however, the problem of determining which groups and individuals are tribal is both subtle and complex. Because it concerns economic interests and the size and location of voting blocs, the question of who are members of Scheduled Tribes rather than Backward Classes (see Glossary) or Scheduled Castes (see Glossary) is often controversial. The apparently wide fluctuation in estimates of South Asia's tribal population through the twentieth century gives a sense of how unclear the distinction between tribal and non-tribal can be. India's 1931 census enumerated 22 million tribal people, in 1941 only 10 million were counted, but by 1961 some 30 million and in 1991 nearly 68 million tribal members were included.

Census Criteria

The differences among the figures reflect changing census criteria and the economic incentives individuals have to maintain or reject classification as a tribal member. These gyrations of census data serve to underline the complex relationship between caste and tribe. Although, in theory, these terms represent different ways of life and ideal types, in reality they stand for a continuum of social groups. In areas of substantial contact between tribes and castes, social and cultural pressures have often tended to move tribes in the direction of becoming castes over a period of years. Tribal peoples with ambitions for social advancement in Indian society at large have tried to gain the classification of caste for their tribes; such efforts conform to the ancient Indian traditions of caste mobility (see Caste and Class, ch. 5). Where tribal leaders prospered, they could hire Brahman priests to construct credible pedigrees and thereby join reasonably high-status castes.

On occasion, an entire tribe or part of a tribe joined a Hindu sect and thus entered the caste system en masse. If a specific tribe engaged in practices that Hindus deemed polluting, the tribe's status when it was assimilated into the caste hierarchy would be affected. Since independence, however, the special benefits available to Scheduled Tribes have convinced many groups, even Hindus and Muslims, that they will enjoy greater advantages if so designated. The schedule gives tribal people incentives to maintain their identity. By the same token, the schedule also includes a number of groups whose "tribal" status, in cultural terms, is dubious at best; in various districts, the list includes Muslims and a congeries of Hindu castes whose main claim seems to be their ability to deliver votes to the party that arranges their listing among the Scheduled Tribes.

Tribal Identity or Caste

A number of traits have customarily been seen as establishing tribal rather than caste identity. These include language, social organization, religious affiliation, economic patterns, geographic location, and self-identification. Recognized tribes typically live in hilly regions somewhat remote from caste settlements; they generally speak a language recognized as tribal. Unlike castes, which are part of a complex and interrelated local economic exchange system, tribes tend to form self-sufficient economic units. Often they practice swidden farming—clearing a field by slash-and-bum methods, planting it for a number of seasons, and then abandoning it for a lengthy fallow period—rather than the intensive farming typical of most of rural India.

For most tribal people, land-use rights traditionally derive simply from tribal membership. Tribal society tends to be egalitarian, its leadership being based on ties of kinship and personality rather than on hereditary status. Tribes typically consist of segmentary lineages whose extended families provide the basis for social organization and control. Unlike caste religion, which recognizes the hegemony of Brahman priests, tribal religion recognizes no authority outside the tribe. Any of these criteria can be called into question in specific instances. Language is not always an accurate indicator of tribal or caste status.

Loss of Mother Tongue

Especially in regions of mixed population, many tribal groups have lost their mother tongues and simply speak local or regional languages. Linguistic assimilation is an ongoing process of considerable complexity. In the highlands of Orissa, for example, the Bondos—a Mundalanguage-speaking tribe-use their own tongue among themselves. Oriya, however, serves as a lingua franca in dealings with Hindu neighbors. Oriya as a prestige language (in the Bondo view), however, has also supplanted the native tongue as the language of ritual. In parts of Assam, historically divided into warring tribes and villages, increased contact among villagers began during the colonial period and has accelerated since independence. A pidgin Assamese developed while educated tribal members learned Hindi and, in the late twentieth century, English. Self-identification and group loyalty are not unfailing markers of tribal identity either. In the case of stratified tribes, the loyalties of clan, kin, and family may well predominate over those of tribe. In addition, tribes cannot always be viewed as people living apart; the degree of isolation of various tribes has varied tremendously.

The Gonds, Santals, and Bhils traditionally have dominated the regions in which they have lived. Moreover, tribal society is not always more egalitarian than the rest of the rural populace; some of the larger tribes, such as the Gonds, are highly stratified.

Economic and Political Conditions

Most Indian tribes are concentrated in heavily forested areas that combine inaccessibility with limited political or economic significance. Historically, the economy of most tribes was subsistence agriculture or hunting and gathering. Tribal members traded with outsiders for the few necessities they lacked, such as salt and iron. A few local Hindu craftsmen might provide such items as cooking utensils.

The twentieth century, however, has seen far-reaching changes in the relationship between tribals in India and the larger society and, by extension, traditional tribal economies. Improved transportation and communications have brought ever deeper intrusions into tribal lands; merchants and a variety of government policies have involved tribal peoples more thoroughly in the cash economy, although by no means on the most favorable of terms. Large areas fell into the hands of nontribals around

1900, when many regions were opened by the government to homestead-style settlement. Immigrants received free land in return for cultivating it. Tribal people, too, could apply for land titles, although even title to the portion of land they happened to be planting that season could not guarantee their ability to continue swidden cultivation. More important, the notion of permanent, individual ownership of land was foreign to most tribals. Land, if seen in terms of ownership at all, was viewed as a communal resource, free to whoever needed it. By the time tribals accepted the necessity of obtaining formal land titles, they had lost the opportunity to lay claim to lands that might rightfully have been considered theirs.

Generally, tribals were severely disadvantaged in dealing with government officials who granted land titles. Albeit belatedly, the colonial regime realized the necessity of protecting tribals of India from the predations of outsiders and prohibited the sale of tribal lands. Although an important loophole in the form of land leases was left open, tribes made some gains in the mid-twentieth century. Despite considerable obstruction by local police and land officials, who were slow to delineate tribal holdings and slower still to offer police protection, some land was returned to tribal peoples.

Erosion of Tribal Gains

In the 1970s, the gains tribal peoples had made in earlier decades were eroded in many regions, especially in central India. Migration into tribal lands increased dramatically, and the deadly combination of constabulary and revenue officers uninterested in tribal welfare and sophisticated nontribals willing and able to bribe local officials was sufficient to deprive many tribals of their landholdings. The means of subverting protective legislation were legion: local officials could be persuaded to ignore land acquisition by non-tribal people, alter land registry records, lease plots of land for short periods and then simply refuse to relinquish them, or induce tribal members to become indebted and attach their lands.

Whatever the means, the result was that many tribal members became landless laborers in the 1960s and 1970s, and regions that a few years earlier had been the exclusive domain of tribes had an increasingly heterogeneous population. Unlike previous eras in which tribal people were shunted into more remote forests, by the 1960s relatively little unoccupied land was available. Government efforts to evict non-tribal members from illegal occupation have proceeded slowly; when evictions occur at all, those ejected are usually members of poor, lower castes.

Skewed Land Ownership

In a 1985 publication, anthropologist Christoph von F-Haimendorf describes this process in Andhra Pradesh: on average only 25 to 33 per cent of the tribal families in such villages had managed to keep even a portion of their holdings. Outsiders had paid about 5 per cent of the market value of the lands they took. Improved communications, roads with motorized traffic, and more frequent government intervention figured in the increased contact that tribal peoples had with outsiders. Tribes fared best where there was little to induce nontribals to settle; cash crops and commercial highways frequently signaled the dismemberment of the tribes.

Merchants have long been a link to the outside world, but in the past they were generally petty traders, and the contact they had with tribal people was transient. By the 1960s and 1970s, the resident non-tribal shopkeeper was a permanent feature of many villages. Shopkeepers often sold liquor on credit, enticing tribal members into debt and into mortgaging their land. In the past, tribes made up shortages before harvest by foraging from the surrounding forest.

The Skylocks

More recently shopkeepers have offered ready credit—with the proviso that loans be repaid in kind with 50 to 100 per cent interest after harvest. Repaying one bag of millet with two bags has set up a cycle of indebtedness from which many have been unable to break loose.

The possibility of cultivators growing a profitable cash crop, such as cotton or castor-oil plants, continues to draw merchants into tribal areas. Non-tribal traders frequently establish an extensive network of relatives and associates as shopkeepers to serve as agents in a number of villages. Cultivators who grow a cash crop often sell to the same merchants, who provide consumption credit throughout the year. The credit carries a high-interest price tag, whereas the tribal peoples' crops are bought at a fraction of the market rate. Cash crops offer a further disadvantage in that they decrease the supply of available foodstuffs and increase tribal dependence on economic forces beyond their control.

This transformation has meant a decline in both the tribes' security and their standard of living. In previous generations, families might have purchased silver jewelry as a form of security; contemporary tribal people are more likely to buy minor consumer goods. Whereas jewelry could serve as collateral in critical emergencies, current purchases simply increase indebtedness. In areas where gathering forest products is remunerative, merchants exchange their products for tribal labour. Indebtedness is so extensive that although such transactions are illegal, traders sometimes "sell" their debtors to other merchants, much like indentured servants.

Acclimatisation

Some tribes learnt the ways the modern world quickly and have managed to hold their own in contacts with outsiders. Some Chenchus, a hunting and gathering tribe of the central hill regions of Andhra Pradesh, have continued to specialize in collecting forest products for sale. Caste Hindus living among them rent land from the Chenchus and pay a portion of the harvest. The Chenchus themselves have responded unenthusiastically to government efforts to induce them to take up farming. Their relationship to Non-tribal people has been one of symbiosis, although there were indications in the early 1980s that other groups were beginning to compete with the Chenchus in gathering forest products.

A large paper mill was cutting bamboo in their territory in a manner that did not allow regeneration, and two groups had begun to collect for sale the same products the Chenchus sell. Dalits settled among them with the help of the Chenchus and learned agriculture from them. The nomadic Banjara herders who graze their cattle in the forest also have been allotted land there. The Chenchus have a certain advantage in dealing with caste Hindus; because of their long association with Hindu hermits and their refusal to eat beef, they are considered an unpolluted caste. Other tribes, particularly in South India, have cultural practices that are offensive to Hindus and, when they are assimilated, are often considered Dalits.

The Clever Ones

Non-tribals, through political jockeying, have managed to gain legal tribal status, that is, to be listed as a Scheduled Tribe. This was a blow for some tribes. The Gonds of Andhra Pradesh effectively lost their only advantage in trying to protect their lands when the Banjaras, a group that had been settling in Gond territory, were classified as a Scheduled Tribe in 1977. Their newly acquired tribal status made the Banjaras eligible to acquire Gond land "legally" and to compete with Gonds for reserved political seats, places in education institutions, and other benefits. Because the Banjaras are not scheduled in neighboring Maharashtra, there has been an influx of Banjara emigrants from that state into Andhra Pradesh in search of better opportunities.

Non-tribals could not make much headway in Himalayan foothills. Tribes in the Himalayan foothills have not been as hard-pressed by the intrusions of nontribals. Historically, their political status was always distinct from the rest of India. Until the British colonial period, there was little effective control by any of the empires centered in peninsular India; the region was populated by autonomous feuding tribes. The British, in efforts to protect the sensitive northeast frontier, followed a policy dubbed the "Inner Line"; non-tribal people were allowed into the areas only with special permission. Post-independence governments have continued the policy, protecting the Himalayan tribes as part of the strategy to secure the border with China.

This policy has generally saved the northern tribes from the kind of exploitation that those elsewhere in South Asia have suffered. In Arunachal Pradesh (formerly part of the North-East Frontier Agency), for example, tribal members control commerce and most lower-level administrative posts. Government construction projects in the region have provided tribes with a significant source of cash-both for setting up businesses and for providing paying customers.

Progress through Education

Some tribes have made rapid progress through the education system. Instruction was begun in Assamese but was eventually changed to Hindi; by the early 1980s, English was taught at most levels. Both education and the increase in ready cash from government spending have permitted tribal people a significant measure of social mobility. The role of early missionaries in providing education was also crucial in Assam.

Government policies on forest reserves have affected tribal peoples profoundly. Wherever the state has chosen to exploit forests, it has seriously undermined the tribes' way of life. Government efforts to reserve forests have precipitated armed (if futile) resistance on the part of the tribal peoples involved. Intensive exploitation of forests has often meant allowing outsiders to cut large areas of trees (while the original tribal inhabitants were restricted from cutting), and ultimately replacing mixed forests capable of sustaining tribal life with single product plantations.

Bribes

Where forests are reserved, nontribals have proved far more sophisticated than their forest counterparts at bribing the necessary local officials to secure effective (if extralegal) use of forestlands. The system of bribing local officials charged with enforcing the reserves is so well established that the rates of bribery are reasonably fixed (by the number of plows a farmer uses or the amount of grain harvested). Tribal people often end up doing unpaid work for Hindus simply because a caste Hindu, who has paid the requisite bribe, can at least ensure a tribal member that he or she will not be evicted from forestlands. The final irony, notes von-Haimendorf, is that the swidden cultivation many tribes practiced had maintained South Asia's forests, whereas the intensive cultivating and commercial interests that replaced the tribal way of life have destroyed the forests.

Extending the system of primary education into tribal areas and reserving places for tribal children in middle and high schools and higher education institutions are central to government policy, but efforts to improve a tribe's educational status have had mixed results. Recruitment of qualified teachers and determination of the appropriate language of instruction also remain troublesome. Commission after commission on the "language question" has called for instruction, at least at the primary level, in the students' native tongue. In some regions, tribal children entering school must begin by learning the official regional language, often one completely unrelated to their tribal tongue.

The Gonds of Andhra

The experiences of the Gonds of Andhra Pradesh provide an example. Primary schooling began there in the 1940s and 1950s. The government selected a group of Gonds who had managed to become semiliterate in Telugu and taught them the basics of written script. These individuals became teachers who taught in Gondi, and their efforts enjoyed a measure of success until the 1970s, when state policy demanded instruction in Telugu. The switch in the language of instruction both made the Gond teachers superfluous because they could not teach in Telugu and also presented the government with the problem of finding reasonably qualified teachers willing to teach in outlying tribal schools.

The commitment of tribes to acquiring a formal education for their children varies considerably. Tribes differ in the extent to which they view education positively. Gonds and Pardhans, two groups in the central hill region, are a case in point. The Gonds are cultivators, and they frequently are reluctant to send their children to school, needing them, they say, to work in the fields. The Pardhans were traditionally bards and ritual specialists, and they have taken to education with enthusiasm. The effectiveness of educational policy likewise varies by region.

Effects of Education

In those parts of the northeast where tribes have generally been spared the wholesale onslaught of outsiders, schooling has helped tribal people to secure political and economic benefits. The education system there has provided a corps of highly trained tribal members in the professions and high-ranking administrative posts. Many tribal schools are plagued by high dropout rates. Children attend for the first three to four years of primary school and gain a smattering of knowledge, only to lapse into illiteracy later. Few who enter continue up to the tenth grade; of those who do, few manage to finish high school. Therefore, very few are eligible to attend institutions of higher education, where the high rate of attrition continues.

4

Characterization of Socio-economic Factors in Western Himalaya : Perspectives and Conservation Implications

Abstract

This component of the Himalayan Biodiversity Project is aimed at characterising the demography and land use patterns of the Western Himalayan states of Himachal Pradesh and Uttaranchal, and to detect socio-economic trends and pointers that would assist in developing conservation plans for the region. We collected data on 44 demographic parameters and 20 land use variables (primary and derived), at various administrative levels, from government agencies; these data were then organized into standardized databases using metadata standards.

A total of 21 districts, 152 tehsils, 167 blocks, and 32,490 villages were covered. The data were subjected to various statistical analyses under three phases: Profile Analysis (summary statistics and graphical exploration), Pattern Analysis (factor analysis and hierarchical clustering), and Correlation Analysis (rank correlations and association measures).

Land Use Profiles

We analyzed and discussed demographic and land use profiles for the states of Himachal Pradesh and Uttaranchal, at various levels ranging from villages to districts. In demography, both the Himalayan states showed identical patterns. However, the land use profiles differed significantly between the two states, particularly the forest cover percentage and decennial change in forest cover. Interestingly, the percentage area under PA network was same for both the states, despite Uttaranchal having higher forest cover.

The district and tehsil level analysis of demography revealed that localities along the foothills of Himalaya have the highest population density in the region and these districts also recorded maximum population growth rate. This was mainly due to the spillover effect of the demography and ecology of the adjacent Indo-Gangetic Plains. While several districts of Uttaranchal showed negative growth in forest cover from 1991 to 1999, those of Himachal Pradesh (barring Kullu) recorded a marginal increase. Other conservation implications of these demographic and land use profiles have been discussed in the paper.

Demographic Parameters

Factor analysis of demographic parameters extracted three major population categories (factors): capital sector workers, land-tilling work force, and mining workers. The second factor, where proportion-of farmers was inversely related to that of marginal laborers, indicated to the unequal distribution of operational land holdings in the hill tracts. Using-these factors, we identified demographically homogeneous clusters of districts to be used in further spatial analyses.

Similarly, we factored the land use variables to derive resource-use categories. Three factors were identified:

hydrological and agronomical land uses, and natural vegetation cover. These factors were then employed to identify clusters of districts that are homogeneous in their land use profiles. We did several correlation analyses to study the interactions among demographic and land use factors. We found that urbanization influenced the relationship of population density and annual population growth rate to proportions of different workforce. There was also an evidence of gender discrimination in socio-cultural milieu of the Himalayan societies, as population of natural resource users was inversely associated with sex ratio at the village level.

We also found that there was an inverse relationship between forest cover percentage and decennial change in forest cover, pointing to the fact that most of the recent observed decline in forest cover may be limited to open village forests. We also found that population density was not significantly related to the actual forest cover; we suspect that this could be due to sampling artifact, as only data on both space and time would possibly reveal the cause-and-effect relationship. In the end, we discuss conservation significance of these findings with an emphasis on socio-economic constraints in developing long-term conservation plans for the Western Himalaya.

"LIVESTOCK MANAGEMENT IN THE HIMALAYAN ENVIRONMENT: CONSTRAINTS, OPPORTUNITIES AND STRATEGIES WITH REFERENCE TO INDIA"- Mtn-Forum On-Line Library Document

Chander, M. and P.C. Harbola, 1996. "Livestock Management in the Himalayan Environment: Constraints, Opportunities and Strategies with Reference to India" in Asian Livestock. Regional Animal Production and Health Commission for Asia and the Pacific, FAO, Bangkok, Thailand, July issue. pp. 78-84.

Introduction

The Himalaya constitutes a unique geographical and geological entity comprising a diverse social, cultural, agro-economic and environmental setup. The Himalayan region in India occupies approximately 591 thousand sq. km. of geographical area (18 per cent of India) inhabited by 6 per cent of Indian population and regulates climate of the entire Indian sub-continent. Thus, the Himalaya, on account of its vastness and distinct entity, biological, social and ethical diversity and problems of the region having far reaching local, regional, and even global consequences, deserves priority for action-planned in ways radically different from those being adopted elsewhere. *(Quote: Shri Kamal Nath, Minister, Environment and Forests, GOI, in his Foreword Message to Action Plan for Himalaya, Himavikas Occasional Publication No. 2, G.B. Pant Institute of Himalayan Environment and Development, Kosi, Abnorn. India, 1992.)*

Problems of Himalayan Regions

The Himalayan regions in India are characterized by all the problems associated usually with the mountainous environment of developing countries, i.e., low availability of cultivable land, lack of diversity in economic activities, low accessibility, low productivity, meager infrastructure, inadequate employment opportunities, an alarming rate of out-migration of skilled personnel, and low level of social and political articulation. Thus, the sustainable agricultural and livestock development is the real challenge in this fragile mountainous environment.

The problems of Himalayan region are as vast and complex as it is itself. The problems are interrelated and part of the Himalayan environment on the whole. It is very difficult, thus, to analyze the constraints and opportunities of livestock management in isolation yet an attempt has been made in this study to focus on constraints in livestock management in general, with emphasis on overcoming these constraints in context of the broader Himalayan environment. The Himalayan economy is basically agro-pastoral and tile dependence on livestock in Himalayan farming system is so much so that it is almost integral part of it. Of the 449 million domestic animals in the country, nearly 9 per cent are in the Himalayan region. The land holdings being very small, these livestock supplement the income and are considered to constitute capital asset. In general, within the framework of mixed farming, these livestock provide wool, meat, milk, hides, skins, motive-power for agricultural operations and transport, manure, etc., and are of immense economical value to the farmers.

Role of Livestock

In addition, livestock provide gainful employment to a large section of population all the year round. Besides, the bulk of the power requirement in the hilly terraced fields is not of much use in this region. If we look at the various livestock species found in the region, cattle are most prominent (47.5 per cent), followed by buffaloes (12.3 per cent), goats (15.9 per cent), and sheep (10.4 per cent). Generally, cattle, goats and sheep constitute an important livestock wealth in western and central Himalaya; pig and poultry in eastern Himalaya, and ovine in alpine zones. Yaks are also reared in alpine areas. Equines are essential for transportation as other means of transportation are meager and also beyond the reach of resource-poor farmers.

There has not been any significant progress as far as the productivity improvement of livestock is concerned in most of the region. Moreover, the diversification in activities involving livestock is also missing. However, any programme of developing the region has to necessarily include the development of livestock. It is important to have an overview of livestock scenario in the whole of Indian Himalaya followed by the constraints in maintaining them in sustainable way then finally on suggesting an effective and practical strategy to overcome the constraints towards the sustainable livestock management.

The region comprises the states of Jammu and Kashmir, Himachal Pradesh, Sikkim, Arunachal Pradesh, Nagaland, Manipur, Mizoram, Tripura, Meghalaya an a part of Assam, along with eight hilly districts of Uttar Pradesh and one district of West Bengal. Its vastness can be gauged from the fact that it is 2,800 km in length and 220 to 300 km wide. The problems relating to agriculture in general, and livestock in particular, have been handled in these states separately as per their administrative convenience and other considerations. Therefore, the constraints are common to the environment irrespective of the situation prevailing in individual state.

Hill Farming

The hill farming system is by and large self-contained; with land, livestock and people being interrelated. Any intervention in one component would mean direct hearing on other components. This calls for a good understanding of the hill farming system or Himalayan environment for that matter in totality be re initiating any developmental intervention. This lack of perception coupled with disregard for the "mountain specificity" has resulted in the failure of many developmental interventions made so far. For example, the cross-breeding programmes for genetic improvement of cattle did not meet much success as the necessary infrastructural support to maintain cross-breed cattle is inadequate (also the acute scarcity of fodder and feed affected these ambitious programmes).

Some of these programmes were sponsored by the European countries with the generous funding, but the desired results eluded the sponsors in most of the cases mainly because the livestock component was seen in isolation and also because of inadequate follow-up efforts coupled with lack of people's participation in these foreign aid oriented projects. Mountainous environment creates difficulties for planned development and thus, offer tremendous challenges to government planners in their attempts to institute rational and efficient programmes for development. The important mountain characteristics or conditions which separate mountain regions from other areas may be called mountain specificity like inaccessibility, heterogeneity, fragility, marginality, and varied human adaptation mechanisms. These mountain characteristics are inter-related and have both biophysical and socio-economic dimensions, and they exhibit considerable variations within the region.

Here, it would be very relevant to quote the example of Uttarakhand region (eight hilly districts of Uttar Pradesh) which has not progressed well vis-a-vis neighbouring mountainous state Himachal Pradesh (H.P.) in the matters of agricultural development in general, and livestock management, in particular. The reason being the plans for Uttarakhand are conceived and finalized at state capital situated in Lucknow (in plains) far away from the hill regions unlike in H.P. Moreover, H.P. could make considerable and rapid progress in livestock sector in particular only after its inception as a state.

Backwardness in Livestock Sector

Therefore, the backwardness in livestock sector is quite obvious in these districts of U.P. The larger policy shift and institutional arrangements at top level are necessary to attain major as well as minor developmental gains. The logical implication, therefore, would be that the mountain specificity can be taken care of effectively only when the policy planning is done after having the micro-view of the area at the grass-roots level for which the development is conceived. Unfortunately, the lessons are rarely learnt mainly due to unwillingness or inability to learn from the previous developmental experiences of government and donor-funded projects.

This lesson of learning exercise is crucial for an effective planning to take place. The experiences can be shared mutually between the state, and even between the countries falling in Himalayan region towards the sustainable development. With the growing global concern for ecology, bio-diversity and sustainable development, the Himalayan environment and development calls for greater attention. The Himalayan environment should be recognized by all concerned as a priority so as to protect it from any Further degradation. Livestock, being an integral component of the Himalayan environment, need even more attention from all the sectors of development. The agri-research institutions, agricultural universities, state animal husbandry departments, cooperative, dairy federations, livestock feed industries, fodder development departments, forest departments, Panchayat Raj Institutions, NGOs and private entrepreneurs, all need to understand the real issues and ground realities having effect oil livestock management and, thus, need to jointly discuss and act towards the sustainable livestock management.

Livestock Management

The management constraints can also be discussed in context of the major livestock management considerations like breeding, feeding, management marketing and disease control under this unique environment. Since cattle constitute the most of livestock in the region, it would be appropriate to discuss in detail about the constraints with respect to cattle development. The productivity of cattle is very low in general and the reason attributed is poor productivity potential of these hill animals. Secondly, the acute shortage of fodder and feed, thus, lack of balance rations in adequate quantities sufficient enough to meet the requirements of higher yields.

It is said that nutrition counts up to 70 per cent towards productivity of livestock. In that case, the poor productivity of hill cattle is obvious, given the fodder and feed shortages in tile region. These cattle mostly depend oil grazing in forest lands which again are poorly managed resulting in their poor productivity. There has to be a well defined location-specific breeding policy for cattle in the region. So far, the breeding policy has been largely governed by the donor/aid agency's dictation. At the moment; no specific policy is being followed about genetic improvement in the entire Himalayan region of India. The indigenous cattle are now being extolled for tile reasons of sustainability, their low production potential notwithstanding.

No Upgradation

Under pressure from environmental groups and organic agricultural movements, the government departments are compelled to issue circulars about conserving the indigenous stock but tile sustained efforts to upgrade their potential by selective breeding are lacking. As such, no clear direction is visible as far as genetic potential improvement/upliftment is concerned. Proper housing providing adequate aeration, sunlight, and cleanliness is important to protect livestock from health risks and to obtain better production from them. Despite the importance of proper housing for better production performance, this does not seem to come in priority as the livestock are by and large kept under dark, closed, and unhygienic conditions. The veterinary services are very poor in most of (lie region, mostly lacking the necessary staff and infrastructural support. Fortunately, the indigenous cattle are well acclimatized to the poor sanitary conditions, inadequate feed and fodder availability, harsh and often erratic climatic conditions, and poor housing, etc.

Poor Veterinary Services

The poor veterinary services do not affect much the productivity of hill livestock but this is a very discouraging or limiting factor towards crossbreeding programmes as the cross-breeds need intensive veterinary health care. The proper housing, feeding, and adequate health care are prerequisite for a successful cross-breeding programme though lack of these is also not justified for indigenous cattle. An effective system of disease diagnosis, monitoring, and control is required within the easy access of the livestock owners. The veterinary dispensaries well equipped with the facilities like vaccination, disease diagnosis, A.I. and other infrastructure need to be strengthened.

Diversification Involving Livestock

Along with cattle, other livestock species hold great potential though not practiced much for various reasons in most of the region. For example, Angora and broiler rabbit rearing is one very good small animal option to be encouraged. Himachal Pradesh's farmers have proved it that the Angora can greatly supplement or even independently boost the income. Besides wool and fur, rabbits can produce the maximum amount of meat per unit area compared to any other species of livestock. Also, rabbits have simple housing and feeding requirements particularly suited to backyard production.

But this success story from H.P. has failed to spread in other Himalayan states probably due to lack of governmental support and lack of effective extension mechanisms. Similarly, poultry farming can be a viable alternative in the lower altitudes like valleys. Currently, the broilers and eggs are transported into hilly regions from the plains, whereas, the local requirements can easily be met by local production, if incentives and technical know-how is provided.

Pashmina goats can be reared at higher altitudes but a remunerative market has to be developed first by encouraging Pashmina-based units in the region. So long, there is no remunerative market locally available for livestock products, it is very difficult to encourage diversification involving livestock. Many farmers would come forward for Angora and broiler rabbit rearing, Pashmina goat rearing, poultry and piggery, etc.; if local markets and processing facilities are developed for the products and by-products of these species.

R & D Infrastructure

A strong R & D infrastructure is needed for livestock development in the different agro-climatic zones in the region. More than this, strong livestock extension mechanisms for timely, relevant and need-based transfer of technology to the farmers are required. Whatever amount any government may spend on research but it would go in vain if concurrent efforts are not made to strengthen to the extension services. Almost the entire Himalayan region in India lacks efficient extension services making inaccessible the much needed developmental advice to the farmers.

Sheep and Goat Breeding

Sheep and goat breeding is being discouraged for the protection of forests and soil conservation. This does not mean that sheep and goats are to be totally eliminated from the production system. Still there are areas where sheep and goats can be reared profitably without any adverse effect on environment provided the relevant recommendations and package of practices are followed. It is a matter of research and insight where and how the sheep and goats can be reared sustainably. For this, research institutions dealing with sheep and goats may come to the rescue of those farmers interested in rearing them.

After touching the issues like breeding, feeding management, disease control and diversification involving livestock, the other constraints hindering the livestock management necessitates some discussion. Tile administrative, political, economic, social, technical, environmental, geographical and ecological constraints have a strong bearing on the livestock management. The very first need is of strong and sincere political will on the part of the government for the livestock development. A Livestock Development Authority may be created for the entire region which would comprise the experts drawn from various disciplines as also from the different zones within the region.

Livestock Sector

A greater allocation may be made by the Central Government as well as by the respective state governments for livestock sector. The integrated efforts of the governments, NGOs, cooperatives, private sectors, entrepreneurs and innovative individuals, are required most to manage the livestock sector. A greater coordination between different development departments within the states, across the states and also across the region is required with the aim of effective livestock management under the situations prevailing in the region.

The research institutions are often busy with tile research activity, yet they need to have a closer interaction with the farming communities to know the problems at grass-roots level. In doing so, the assistance of the genuine local NGOs may be taken as the NGOs often maintain close contacts with the farmers. Accordingly, the research priorities may be oriented towards solving the real problems the livestock owners face in field conditions. The role of NGOs, thus, needs to be' recognized in the livestock management.

Overcoming the Constraints

Therefore, the strategy to overcome the constraints discussed above should include following as necessary conditions for the sustainable livestock management :

1. Greater coordination and collaboration, is required between different development departments within the state, across the states, and also across the region. For example, there is a need to collaborate more intensively with the international institutions like International Centre for Integrated Mountain Development (ICIMOD), Nepal. At the moment, such collaborations are very limited as only a very few organizations are involved in collaborative research and development work in the area of livestock management.

2. A sound strategy is required for the genetic improvement/up-grading of hill livestock. It is by no means a simple task as the ability of local livestock to perform better under local conditions can not be under-estimated for long. An ideal mix of both the considerations has to be taken care of before arriving at any logical conclusion over the issue of productivity improvement that involve internal or external genetic interventions.

3. Higher productivity of livestock can be maintained only when there is sustained supply of adequate feed and fodder. Nutritive values of grasses and fodder also need to be improved by introducing improved species suitable for the region. The hilly regions of Kerala, and to some extent H.P. as well, have progressed very well in the area of livestock management including fodder availability. Such experiences can be shared fruitfully. Since most of livestock depend on grazing in the forest areas and grazing lands, the productivity of such areas needs to be improved by proper management in collaboration with the forest departments. The infra-structural development along with the massive transfer of technology is required to promote' scientific feeding practices. It is the experience of IVRI in the areas of Kumaon hills that the farmers were totally ignorant of balance feeding practices.

4. More well-equipped veterinary dispensaries with A.I. facilities are needed in the rural areas; and wherever required, mobile veterinary dispensaries may also be put into service with the aim of providing the health care services as close as possible to the livestock owners.

5. The governmental support in the forms of credit facilities to purchase better livestock, inputs, etc. could help greatly to promote livestock management including diversification in activities. The infrastructural facilities development should form the priority it for the government to facilitate speedy livestock development. The revamping of centrally-sponsored *"Krishi Vigyan Kendras"* (KVKs) is also required to ensure the skill oriented trainings to the farmers. The state governments should have their own extension mechanisms on the lines of KVKs with links with the credit institutions.

6. Diversification in activities involving livestock should be encouraged by providing technical, financial and marketing supports. The livestock based units including processing facilities in the vicinity of production centres will greatly help boost livestock sector, The respective governments and also a nodal agency at central level specifically created to monitor the progress with regards to diversification will help greatly in this matter. The rural youth can be productively involved in Angora or broiler rabbitary, piggery, etc. Diversification is never an easy process. Sincere efforts can bring in the desired results.

7. Promotion of cooperatives will help eliminate the middlemen. The success of horticultural cooperatives in Himachal Pradesh and milk cooperatives of Gujarat may be viewed positively for livestock sector in the region as well. This is particularly required for milk marketing. In the beginning, the cooperatives may be provided

adequate financial back-up so that these can stand on their own after sometimes. The investment in cooperative sector will have a positive impact in the long run, not only in pure economic terms but also in socio-cultural development context as well.

8. Women in mountainous regions do most of livestock management yet their role is not recognized officially adding to their drudgery. These women need to be made aware of and educated oil livestock management practices by, involving them more in formal livestock development programmes. The women's dairy co-operatives which-are being fanned in various areas of the region are good example of a development intervention with greater role of rural women. Women need to be integrated in all the activities relating to livestock management at formal level in recognition of their contribution at informal level.

9. The NGOs are now mushrooming in the entire region and mostly claim they are for protecting the fast degrading environment. The services of carefully chosen NGOs may be channelized for livestock management by providing them financial and logistic supports. The NGO-NGO interaction is very much required for better results in livestock sector. Some NGOs are doing really very good work for livestock development in hills of U.P. and in some North-eastern states as well. Their works need to be appreciated, encouraged and supported by all possible means of the governments.

10. An efficient extension setup, well equipped with technical know-how and solid infrastructural back-up, is required to cater to the local needs of different agro-climatic zones. The extension machinery has to be an ideal bridge between the research/development institutions and the beneficiaries. At the moment, lack of all efficient extension mechanism is a big constraint coming in the way of livestock management.

11. Finally, no programme would meet success if people's participation is not ensured in the programme. The livestock owners/beneficiaries need to be taken in confidence before initiating any intervention for their betterment. Many programmes have failed in the past only because of the non-participation of the people. Thus, the efforts should be made to make the livestock management programmes people centered, participatory in nature.

Conclusion

In this study, an attempt has been made to describe the situation involving the constraints in livestock management and a few points for consideration have also been offered. The efforts have not been made towards providing the data on livestock scenario but emphasis has been oil describing the situation. Moreover, the lack of data on livestock sector in this region is one big constraint in itself which needs to be looked into if systematic efforts are to be made towards the livestock management.

Browsing Classification: Agriculture: Animal Husbandry: Asia Pacific: India Agriculture: Elevage Agricultura: Ganaderia.

Citation: Chander, M. and P.C. Harbola. 1996. *"Livestock Management in the Himalayan Environment: Constraints, Opportunities and Strategies with Reference to India"* in Asian Livestock. Regional Animal Production and Health Commission for Asia and the Pacific, FAO. Bangkok, Thailand. July issue. pp. 78-84.

5

Tibet and Development : A Personal View

Summary of a talk given in London on 30 October 2003 by Phuntsog Wangyal*

With the Chinese coming into Tibet, its development became an important subject, often highlighted to draw attention to the 'backwardness' of Tibet and of its economic system. Development of Tibet is a subject much debated among the Tibetans and often this subject is seen primarily in its political context. In the past history of Tibet you can find references to the development of religion, Buddhism or a particular philosophy but hardly any reference to the development of Tibet, or its economy.

Broadly speaking the word "development" is accepted to mean an action or activity that causes a situation to change for the better or progress. It is also believed, in the consensus of opinion, that fostering economic development and prosperity is a 'positive' action, as almost any leader would take pride when his/her country's economy is performing strongly.

---

*Member of the Board of Trustees of Tibet Foundation, former London representative of His Holiness the Dalai Lama at the Office of Tibet, member of the Tibetan Assembly, and director of Tibet Foundation.

In the Past

Tibet had a highly developed spiritual tradition. Its civilization, largely spiritual—Buddhist and Bon (Tibet's native religion)—was well documented. By the beginning of the twentieth century Tibet had one of the largest collections of literature on Buddhism. All those who read about Tibet happily acknowledged the uniqueness and richness of its religious and cultural traditions. Yet there was hardly any substantial writing on social and economic subjects.

Many wondered how and why life in Tibet remained unchanged, unaffected by the new social, economic and political ideas that transformed life in many other parts of the world. Looking back, one could see that Tibetan people always took great pride in being Tibetan. At the same time they were also seen to be open to new ideas. They had, no difficulty in adopting Buddhist philosophy from India and many other cultural habits and customs from China and Mongolia.

Chinese Influence

Although the Buddhist religion remained the driving force in their everyday lives, Tibetans were still very creative and progressive in their attitude towards, the social and artistic culture of their neighbour in the east, China. Most noticeably, the custom of drinking tea, eating vegetables (many names of vegetables are still in the Chinese language), wearing silk, and aspects of astrology, art and musical traditions are some examples of many influences happily borrowed from China. Tibet on the whole remained highly isolated. The high attitude (averaging over 4,000 metres, or 13,000 feet, above sea level), the low population of just 6 million, (in all ethnic Tibetan areas) and its vast size (almost that of Western Europe) contributed much to Tibet's isolation from the rest of the world.

This, in turn, no doubt contributed to the Tibetan people's ability to maintain the uniqueness of their culture and the diversity of many different dialects and regional characters. Economically, Tibet was backward. When the West was going through-successive explosions of development in science and technology, Tibet was going deeper inwards, which they proudly called 'understanding of the inner world' of spirituality. It became the dominant philosophical view that life in this world is far shorter than the life in the next world. Pursuing a contented life now for the sake of long-lasting happiness in the future was the dominant—and much respected—ideology. Therefore in economic terms, there was hardly any development: no electricity, no motor vehicles, no telephone and no modern machinery of any kind.

Yet the Tibetans, having hardly ever seen any other way of life, felt fully content with their life and missed nothing but "enlightenment". It was a common practice that many spent a large part of their time in religious activities such as chanting mantras, circum-ambulating temples, saying prayers, turning prayer wheels etc.—activities which the Chinese called "economically unproductive". A quarter of the whole population was known to be either monks or nuns. Perhaps Tibetan people at that time were one of the most 'religious' people on earth. Religion was their most precious possession and they were happy to practise it. They were living in a different world, where "simple living-high thinking" dominated their lives.

Since 1951

That situation changed completely in 1951, when Tibet officially became a part of China, a new reality that the Tibetans had to face with hardships never witnessed by their forefathers. Old Tibet, just like a dream, was transformed into a communist state. The process of this transformation was revolutionary, and in that process the

Tibetan people found themselves in a position where they had no-or very little-influence in shaping the future of their country. Every important decision was made in Beijing, not Lhasa. It was no longer religious but political and economic considerations that mattered most.

The right or wrong of this historical transformation is not a subject under discussion in this study. Today we would rather look at the Tibetan people in this changed situation and whether Tibetans (especially those living under the Chinese rule) wished and deserved economic prosperity.

The New Generation

During the fifty-odd years of Chinese rule, a whole new generation of Tibetans were born. They were brought up under a system of government strictly shaped by communist ideology and mainly administered by the Chinese. These Tibetans never had the opportunity to learn their own philosophy as their forefathers did for generations. On the other hand, they were very much exposed to a new materialistic world with new ideas of science and technology. A few had some education and training to be cadres helping to run the administration. Many never had any opportunity to study anything, let alone Buddhist philosophy. Yet the situation in Tibet went on changing in accordance with the changes that were taking place in China. In the past decade or so, China has been changing even faster. We began to see a government that, for the first time, adopted a somewhat liberal policy towards economic reforms. Ordinary people began to enjoy, comparatively speaking, some freedom of movement within the country, ownership of private property, more access to education and less [government] control over their family life. In short a system primarily based on the market economy is emerging in China.

Geographically, China is a huge country with a very uneven distribution of population and wealth. In all Tibetan areas, the density of population is low and the level of poverty is extremely high; the level of education is low and the death rate is comparatively high; and above all, the basic infrastructure is extremely poor.

Lack of Manufacturing Sector

There is hardly any manufacturing industry in Tibet. Local economics are almost completely based on agriculture and animal husbandry. The tools and methods used are still very old fashioned. In short, Tibetan areas today are some of the least developed and most backward even from the Chinese government's point of view. It is fair to say that governments at every level are making efforts to improve the situation in Tibet by investing both money and human resources from China. They believe that the key to their success lies on their strategy to develop Tibet's economy.

Lack of Education

Unfortunately, not many Tibetans in Tibet are able to get involved in the process of development. This is by no means either due to their inability or their lack of interest. It is primarily due to a lack of education and/or required skills amongst the Tibetans. In sharp contrast, the Tibetans in exile, with better capital resources and adequate training and education, could make a big difference in advancing their standard of living.

Economic Prosperity

Religious aspiration may still be high on the agenda, yet the general attitude, rightly or wrongly, seems to be changing. Young Tibetans have strong feelings for their religion, more in terms of their identity rather than as their philosophy. No one could or should ignore the fact that they too have desires for a better living standard and economic prosperity. Living in yak hair tents at subzero temperatures and burning cow dung as their fuel may look romantic, but those who live their lives in such poor conditions may have a different view. They are not to be blamed.

The urge for economic prosperity is in fact more noticeable among those Tibetans who live a comparatively better life in India and Nepal. Many young Tibetans, including many monks an nuns, are leaving or trying to leave their homes and monasteries in India for the, West and 'a better life'. The scale of the migration was such that the senior Tibetan Kalon, Samdhong Rinpoche, has seen this as a major problem and announced publicly that Tibetans should refrain from leaving their settlements in India.

Development Schemes

The Central Tibetan Administration has since developed a number of schemes to improve the economy with the hope that Tibetans will stay in their settlements where they have a better chance of sustaining their culture. The Chinese and the Tibetans both agree that the situation in Tibet needs to improve. But they greatly disagree about what needs to change first: the political or economic conditions. The changing political situation in Tibet requires more than just goodwill or sympathy. A major political change in China is necessary. At the moment, such a political change is most unlikely, as the Chinese leadership is fully engaged and occupied with major strategies to advance the economy of the country, and safeguard the unity of their 'motherland'. This shift of policy towards economic development may not be the favourite option for the Tibetans but this itself presents an opportunity for them to develop their education, healthcare and economy.

Tibetans in Business

Although very few in number, some Tibetans are already making great progress. You might be surprised to see some of them engaged in large business ventures, even building three-star hotels, supporting educational and healthcare facilities and managing handicraft and business centres for the rural economy. With the right approach they are now able to do many positive things that they could not have dreamt of in the past.

Role of NGOs

This new situation provides Tibetans in Tibet with some opportunities to take control of their lives, if in very small ways. It provides those of us in the West with opportunities to support them in their efforts to achieve some progress. In the past, there was no meaningful role for a voluntary or non-governmental organisation (NGO) in China. It was the state that initiated new ideas, managed organisations and controlled economic processes. We can see this slowly changing as more national NGOs are being formed and more international NGOs are involved in development works in China.

A Chinese scholar stated, "The role of NGOs in the 21st century will be as significant as the role of the nation state in the 20th century." Having said this, we must note that the Chinese government, at every level controlled by the communist party, is still too strong compared to a 'civil society' that is yet to mature and to NGOs that are still in their infancy. Another factor worth taking into account is that the Chinese government is placing a strong emphasis on "democracy" and the role of "democracy" in development, although many may argue about the substance of their emphasis and their meaning of the word "democracy". The Chinese government lacks trust in Western democracies and they are sceptical of donors' motivations.

Chinese Strategy

Regarding the Chinese Strategy, the aims of the Chinese development policies are unmistakably to bring about greater geographical and economic integration of the country as a whole, including areas such as Tibet. The motivation may not necessarily be the destruction of the local economy or local culture as is often alleged. However, the outcome of these development schemes is bound to have an impact on delicate local issues. If they are inadequately managed, development will cause irreparable damage to irreplaceable Tibetan assets such as its culture and environment. The government's economic policy is directed to enhancing development through massive infrastructure such as communications and school buildings, and providing incentives to local businesses, job creation initiatives and tourism. We could see some degree of consistency in their push in this direction. There is no sign of a major change in current Chinese policies.

Liberal Policies

It is true that many in the West would like to see major changes towards more democracy and more participation by the general public in matters affecting their lives. Although this is not happening fast enough, there is every indication that China seems to be heading in that direction. In relative terms, the current liberal policies are providing new opportunities for those of us trying to help Tibetans in their areas in China. We are all very aware of the fact that many government agencies had national and international NGOs are already working in many Tibetan areas. Financial support given by the Tibetans and their supporters in the West is estimated at around 10 million Yuan per year only (approx. less than £700,000). Their contribution goes largely towards rebuilding and renovating religious sites and small education and healthcare projects.

Working in Tibet is not impossible, but still difficult and sometimes frustrating. Often those who work in Tibet face many obstacles; progress is slow and gradual. In spite of these obstacles, working with whatever facilities are available is a better option and is perhaps the best and most practical approach which one could take the current situation. Thus it is sensible to encourage NGOs' efforts in ensuring that their assistance reaches tile most needy local Tibetans and empowers them to help themselves.

Obstacles

It is equally important to understand the nature of the obstacles many NGOs face in carrying out their work. I would like to present a few, factors that may be contributing to (a) international NGOs not receiving adequate funds, and the poorest people and most remote areas of Tibet not benefiting from potential international aid:

First, although many international NGOs are well aware of the importance of getting local people involved in all aspects of local projects, the authorities seldom consult local Tibetans or meaningfully involve them in projects, especially big projects, which have the potential to greatly impact their lives. This is not necessarily by the design of the Chinese authorities, though it, is often alleged to be so. Primarily it is due to a lack of adequate knowledge and expertise on the part of the local people. In most cases decisions are taken by authorities who are likely to be Chinese, especially at higher levels, who have no special concern for, nor understanding of, the local culture or environment.

Second, donors, especially government and corporate funding bodies, often require projects which return demonstrable results in a short time frame. The deep-rooted needs of Tibet, coupled with the low population density, short summer months, and low levels of physical, social and intellectual infrastructure, make proving tangible results in the time frame sought by these donors extremely difficult. In some cases, particularly those where long-term or smallscale projects are appropriate, providing evidence of these results can be more demanding than actually achieving' the results themselves. This makes centralised large-scale projects appear more successful in getting funds than smaller or more highly devolved projects, and in some instances larger scale intervention becomes the only eligible recipient of support.

Third, governmental and large multinational agencies generally prefer to deal directly with central or provincial governments in China. At this level, ethnic Tibetans are unlikely to be the decision-makers. The more you are able to work with local people in Tibet, the closer you will get to understanding what they actually need. The smaller the scale of your work in a smaller area, the more benefit, you are likely to deliver to the people.

Fourth, although the works of international NGOs are becoming increasingly more acceptable in China, the authorities still see them as foreign agencies trying to import 'western values'—generally believed to be anti-communist, and thus anti-China. President Bush's efforts to establish "the principle of pre-emptive military intervention to protect American interest" and the criticism of the Chinese Western Development Programme as "a deliberate design to destroy Tibetan culture" have not been entirely conducive to fostering the trust which is so desperately lacking.

Thus, the challenge that Tibetans face is enormous. The world in which they live is no longer isolated. The society in which they work is more multi-racial and highly competitive. There is a great demand on limited resources, and competition with an incoming population, often better educated and better informed, is an almost impossible challenge to the Tibetans.

Investment for Development

Tibet needs development: a development of its economy to enable Tibetans to survive in a fast-growing competitive world. Development need's investment: an investment in promoting education, improving healthcare facilities and raising living standards. Tibet needs support that is sustainable and that helps Tibetans to be self-reliant. Many Tibetans show concern about the suitability of large-scale national projects where there is little or no input from the local people.

Sustainable Projects

This must not be confused with regional projects close to local people and local areas. Projects with investment in local areas and with input from local people are proving to be more sustainable and a real help to the people. Investment in small projects is not an attractive venture to private investors, as there is no real financial incentive nor does it bring immediate results for NGOs, who may need to satisfy their donors. But such investment in small projects close to the local community will be an invaluable benefit so the Tibetan people in the long run.

This is a time when the Tibetans have to be innovative, creative, progressive and open to ideas. And this is the time tune that their supporters must realise that the Tibetans are real people and that they have to live not in isolation but in a most competitive and challenging world. A Tibetan saying goes, "getting simple things done now is far better than thinking of doing many great things in the future". This couldn't be truer today. It is my belief that Tibetans are doing what they can in their own small ways under a very difficult situation in Tibet. They are proud of their honest, ethical and trustworthy.

Need for Support

The need for support from all available sources is real. The success of such support to the Tibetans depends on a broad understanding of their needs in Tibet, the driving force behind economic development and what matters most to the Tibetans in their daily lives in their unique situation. Politicising all issues related to Tibet may not necessarily produce right or expected results. The government's trust in NGOs' commitment and the NGOs' reassurance of the non-political nature of their activities are crucial, and their importance must not be underestimated. This leads me to say that there is a greater need of strong partnership between local communities that invite investment, NGOs that provide funds, and governments that supervise their management. To achieve such a partnership may not be easy, but it is an effort worth making. Supporting small projects close to the community may not be most attractive, but such projects are the most valuable to the people in Tibet.

6

Impact of Man on the Mountain Ecosystem

At present, the influence of mountains on the climate, vegetation, animal life, socio-economic conditions and geo-political problems is very little understood. It is also only very recently that increasing attention is being paid to the impact of human activities on ecosystems on mountains in different parts of the world. In India, however, both the aspects of ecology have been sadly neglected so far. In technologically advanced countries, the massive interference by man has caused irreversible changes in the natural balance of mountain ecosystems, which has had a profound influence on the upland and lowland ecology thus affecting human welfare directly and indirectly.

Man and Biosphere

Under the programme Man and Biosphere (MAB), the UNESCO organized a series of conferences to formulate research programmes by different nations. Recently a meeting of the working group No. 6. Impact of human activities on mountain and tundra ecosystems, was organized by UNESCO, from November 20 to 23 1973, at Lillehammer, Norway, where specialists from different nations including India met to discuss this aspect of the problem. In this meeting the school of entomology made a significant contribution.*

This meeting recognized the school of entomology, as a central co-ordinating unit for research on Mountain ecosystem for the Indian region. The Institution has been actively engaged in researches in mountain ecology and the high altitude biology for two workers from of observing at first hand the rapid changes which have been brought about in the mountain ecosystem in the Himalaya due to various human activities such as disastrous deforestation, overgrazing, ruthless hunting of wild animals for fur, wool, musk and other valuable products, construction of large-scale hydro-electric installations, tunnels, drains, impounding of massive volumes of water by gigantic dams, construction of high mountain roads, air-fields, and by the unprecendented influx and settlement of Tibetan refugees. These activities have brought about the total loss of virgin lands, irreversible changes in the natural vegetation cover and massive destruction and impoverishment of flora and fauna, soil erosion, water pollution and silting up of general Himalayan river systems, as well as destructive floods in the sub-Himalayan tracts.

The Change Makers

Inhabitants of the mountains only partly brought about changes. These alterations and interferences have been brought about by man from the plains. In India, man has subjected the Himalayan mountain to intense stress and strain over the last twenty-five hundred years. Though no quantitative investigations have so far been carried out to evaluate the effects of these changes, yet observations on qualitative changes have been periodically made by the workers from the school of entomology.

---

* Dr. M.S. Mani, Professor Eneritus, School of Entomology, represented the Government of India.

Some of these observations are briefly reviewed here with a view to draw specific attention to urgent problems. In these observations attention has been drawn to the ill effects of deforestation, selection of plant species of doubtful suitability for afforestation, over grazing on the mountain slopes and alpine meadows above the timber line, construction of dams, mountaineering and tourism, hunting of wild life and the supersonic flights.

II. IMPACT OF FOOD, FODDER AND FUEL ON THE TERRESTRIAL ECOSYSTEM OF OUTER KUMAUN HIMALAYAS

The forest tree canopy with its undergrowth of shrubs, perennial and annual herbs and the trunk laden with climber, epiphytic orchids, ferns, mosses and lichens together with great stretches of bryophytes and ferns on the ground and the rocks, constitutes the primary source of energy in the mountain ecosystem. So green an expanse should be a tremendous force in the production of biomass.

Man in an effort to produce more food found it easier to extend his fields by clearing the adjoining forest rather than use improved agricultural technology and in so doing the forests overlooking valleys have been denuded in due course. In search of more fodder he lopped the trees and his cattle overgrazed the green herbage and greases. Need for fuel-wood in absence of any alternative source of energy gave further fillip to the denudation process. Besides, many herbs, shrubs and trees of medicinal and minor industrial importance have been continuously under man's ravages.

Systematic Deterioration

On the treks undertaken by the author during 1944-47, 1955,1960-62 and 1974-76 the vegetation of the outer

Kumaun hills was noticed to have undergone progressive deterioration, causing imbalances in the terrestrial ecosystem. In the sub-montance region where thick moist deciduous forest dominated with heavy climbers and impenetrable undergrowth, the forest appears to be thinned down, abundance of lianas no more and the undergrowth seems cleared.

There has been fast decline Adina cordifolia, Terminalia tementosa, T. belerica, edrela toona and Bauhinia variegata. At places shorea robusta, allotus phillippinensis and fcastanopsis tribuloides are practically extinct. This is the case with several perennial herbs and shrubs whose place is fast being taken up by santan camera which has made deep and extensive inroads not only in the sub-montane region but also higher up. It has penetrated the cultivated fields to such an extent that the farmer has abandoned cultivation in desperation. The Australian bug has yet to play its role.

Above the sub-montaine region is the pine belt with a few deciduous tree species followed further up, chiefly on the southern aspect, with pure pine *(Pinus longifolia)*, mostly planted. Cedrela toona once a prominent associate in the lower region has almost disappeared under human and cattle pressure. Pine too has been subject to indispriminate exploitation for fuel and timber or sapped of its growth and strength by deep and extensive resin tapping. Pine has been completely stripped off at many places. In the higher regions pine merges with oak *(Quercus incana)* and finally *Q. incana* dominates mixed with *Q. dilatata* or at higher altitudes with Q. semecarpifolia. Oak supports a rich floor vegetation which on decomposition adds deep layers of humus to the soil. The trunk is fully laden with mosses, ferns and lichens.

Forest Denudation

Because of its great usefulness oak has been subject to great devastation and oak formations have been reduced to oak-scrub in a state of biotic climax. Unlike pine, regeneration in oak is almost none inspite of a large seed out-put. The seeds are all infected with a pest. The forest denudation has brought about disastrous consequences by rendering land unsuitable for agriculture and suitable only for such xerophytic species as *Eupohorbia royaleana*, water springs to dry up, hill-sides to erode etc. not to talk of disappearance of bird life or other once rich fauna. Not only the green mantle as primary producers but also the great array of micro-organisms which play an important role in the terrestrial ecosystem have undergone damage during the course of years.

III. EXTINCT AND THREATENED WILD LIFE IN WESTERN HIMALAYA

In the last few decades, the interior and hitherto inaccessible regions of the western Himalayas have been opened up. It was done out of economic necessity. Due to this, the wild life of the region has been affected very adversely. It has not only suffered on account of its ever shrinking habitat but also by the easy access to the poacher to their otherwise difficult to reach habitat, provided by the recently constructed roads in the region.

About two decades ago the author had the rare good fortune to trek the Western Himalaya for about 40 weeks on foot covering approximately 2500 kms; right from the foot hills to a number of high Himalayan passes between 3000 to 5750 metres. During this period he had observed and taken note of the wild life population of the region. In the last two years he has again visited some of the areas he had visited two decades ago. His impression and observation on the present state of diminished wild life population have compelled him to conclude that the following species of Mammals and Birds endemic to the region may not survive for long unless some thing darastic is done for their conservation.

Mammals

1. The Tiger, *Panthera tigris* (Linnaeus).
2. The Snow Leopard, *Panthera uncia* (Schreber).
3. The Fishing Cat, *Felis viverrina* (Bennett).
4. The Lynx, *Felis lynx*(Linnaeus).
5. The Bharal, *Pseudois nayaur* (Hodgson).
6. The Himalayan Tahr, *Hemitragus jemlahicus* (H. Smith).
7. The Serow, *Capricornis sumatraensis* (Bechstein).
8. The Swamp Deer, *Cervus duvauceli* (Cuvier).
9. The Musk Deer, *Moschus moschiferus* (Linnaeus).

Birds

1. Himalayan Bearded Vulture, *Gypateus barbatus aureus* (Hablizl).
2. Mountain Quail, *Ophrysia superciliosa* (J.E. Gray).

3. Western Horned Pheasant, *Tragopan melanocephalus* (J.E. Gray).
4. Crimson Horned Pheasant, *Tragopan satyra* (Linnaeus).
5. Himalayan Monal Pheasant, *Lophophorus impeganus* (Latham).
6. Chir Pheasant, *Catreus wallichu* (Hardwicke).
7. Great Pied Horbill, *Buceros bicornis homrai* (Hodgson).

IV. HOUSEFLY POPULATIONS AT HIGHER ALTITUDES AND THEIR SENSITIVITY TO INSECTICIDES

Housefly has been the man's companion ever since he came into existence. Perhaps it was there before man appeared as nature's creature. The fifth loving housefly is not always recognized with certainly and it is doubted if the nonmetallic houseflies which occur at various altitudes belong to the same species or that true Musca domestica is absent from certain geographical regions. Deoras and Jandu (1943) believe that the predominent from of housefly in India is Musca nebulo. During our work on the reproductive ecology and insecticide resistance in houseflies, we collected these flies from various parts of India and found that besides *M. nebulo* two other forms of such flies, *M.d. vicina* and *M. sorabans* were prevalent in various proportions in different parts of the country. In general *M.d. vicina* was more abundant than *M. nebulo* at high altitudes.

Biological Requirements

This may be due to certain biological requirements of this form of housefly. Perhaps the higher temperature of the summer months retards its oviposition in the plains where *M. nebulo* becomes predominently abundant. A study of the inheritance of the thoracic stripes in nebulo and vicina showed that the two forms are inter-fertile. It may therefore be more appropriate to consider them as subspecies of *M. domestica* rather than as valid species. Both *M.d. nebulo* and *M.d. vicina* were found to be about equally susceptible to DDT but *M.d. vicina* could develop DDT tolerance much more rapidly than *M.d. nebulo*. This suggests that the use of modern synthetic insecticides may be more hazardous at higher altitudes than in the plains.

Musca sorbans could not be obtained from the hill stations of Chakrata in Uttar Pradesh and Simla in Himachal Pradesh. It needs a temperature of 24° C to 27° C to enter the houses and act as an effective facultative blood sucker (Zinin, 1941). Thus the low temperature found at high altitudes may be a boon for mountain dweller as there is considerable evidence that *M. sorbans* serves as a vector of leprosy and yaws.

V. ECOLOGY AND DISTRIBUTION OF GRASSHOPPERS (ORTHOPTERA: ACHRIDOIDEA) IN ARUNACHAL PRADESH, IN INDIA AND IMPACT OF HUMAN ACTIVITIES ON THEIR ECOLOGY AND DISTRIBUTION

Grasshopper fauna of Arunachal Pradesh has been explored by the zoological survey of India since 1961. Collections of grasshoppers have been made from the inner and outer valleys of Kemeng, Subansiri, Siang, Lohit and Tirap districts of Arunachal Pradesh where the vegetation shows considerable variation from evergreen tropical, subtropical, temperate and subalpine forest regions to alpine in different parts of the year.

As a result of these collections 28 species (excluding five dealt with here up to generic level as their details will be published elsewhere) distributed over 23 genera and three families are reported here for the first time. Notes on the ecology and distribution of these species are given. Under the head habitat and abundance is included data during rainy season. There was a tendency in some of the species to be greener during rainy season whereas dark-brown or black specimens were associated with burnt areas. Specimens from higher altitudes were generally dark and small in size. Seasonal changes in colouration were distinctly marked in *Acrida ezalta*.

The study of ecosystem in relation to its fauna is a complex one and includes the study of change in environment which is the most powerful stimulus for variation in organisms.

In recent years Arunachal Pradesh has received attention due to its strategic position and the policy of the government to develop the neglected and backward regions of the country. These developmental activities, which have involved deforestation, cultivation, road construction, etc. have exerted a profound influence on land and on many elements of the natural environment. The change in environment has affected the animal population in general and grasshoppers in particular.

Intrusion of Man

The intrusion of man into well established complexes of grasshoppers, while cultivating areas of virgin lands, has led to two developments. One is a considerable reduction of both the number of grasshoppers species and the total number of grasshoppers in the newly formed fields. The other development is a contrary one, taking place in the small patches of virgin land, in depressions and in fallows in all of which grasshoppers are forced to concentrate and liable to mass multiplication In favorable years. Karelina (1961) reported same 2000 individuals of chorthippus albomarginatus per square meter in Yakutia and at such densities signs of phase change were noticed. During the present study we noticed that Xenocatantops humilis humilis and Stenocatantops splendens, habitual inhabitants of forest, have been forced to occupy new zones which are quite different from their own natural habitat.

It has also been observed by us that Trilophidia annulata is a typically geophile species but in thick forest regions in foot hills of Arunachal Pradesh it often behaves as a phytophile. All though no heavy infestation of Pantanga succincta was seen in Arunachal Pradesh, there is a possibility of this because of the deforestation. Pagden (1959) reported an outbreak of Patanga in Malaya when the forest was cleared and the area was brought under plow. Other species of grasshopper death in this study are also affected by change in environment and some of these could become permanent pests.

A part from mass breeding and destruction of crops such adaptive displacement of grasshoppers can lead to change in ecosystem, as these insects accelerate the microbial break-down of phytomass. These studies open up certain new avenues like speciation, control measures in the plant protection acclimatization, etc.

VI. SOME OF THE HARMFUL EFFECTS OF HUMAN ACTIVITIES ON THE FISH FAUNA IN THE HIMALAYA

Fishes are an important form of aquatic life in a large number of streams and reservoirs in the Himalaya. Some of the illegal methods of fishing such as overfishing and dynamiting huge bodies of water, poisoning of streams and pools, using unapproved gears of fishing and fishing during close season are briefly touched upon. The introduction of exotic fishes, which are fast breeders, has also been adjudged as an unfavorable event for the well being of the endemic fish stock of Himalaya. Large bodies of impounded waters, meant mainly for hydroelectric generation of power, form a very useful and important abode of fishes but it has been discovered that majority of them are not well managed.

Since care and production of fish is of tertiary importance in these lakes and reservoirs, generation of electricity and irrigation being of primary and secondary importance respectively due consideration to fisheries aspects has not been given. Moreover, fisheries plan was not included by the Hydroelectric Planning Engineers in the blue-print before the construction of these reservoirs. As such the fisheries scientists, connected with these reservoirs for the cause of fisheries which suffers as a result. A request for the creation of fish sanctuaries in the running or standing bodies of water has been reiterated so that they could serve as a source of supply of fish shield to the waterways where fish stock is gradually getting depleted.

We know almost nothing about the physiology of the environment-adaptation of the high altitude species, and this is a fruitful field for future research.

VII. IMPACT OF ANTHOROPOGENOUS WEEDS IN INDOMONTANE ECOSYSTEM

The indomontane flora is influenced by the three groups of factors: climate, soil, mans and his associated factors (anthropic factors). The powerful influence of man with the introduction of intensive farming, large-scale forest clearance and man-made habitats for rehabilitation and settlement projects and for industrialization has brought a radical change in vegetation. Thus, mixed plant communities composed of indigenous and adventive element, plantations of exotics (e.g. Salix, Populus and Castanea in Kashmir, Acacia, Cinchona and Eucalyptus in the Nilgiris, Fcryptomeria in Darjeeling Hills, Tea in North Bengal Hills, Assam, Nilgiris and Malabar, Pinusin south Indian Hills and in Khasi and Jaintia Hills of Meghalaya, and pure associations of adventive weeds (e.g. Eupatorium, Lantana) are seen in the flora. Man and his associated agents (viz. Fire, ballast, grain, shifting cultivation, grazing and migratory animals-cow, sheep, goat and yak, etc.) have played an active role as an intentional or accidental carrier of diaspores.

The spread of diaspores and of light fruits and seeds over long distances in the montane regions was further facilitated by natural means like wind, water and dust-devils and aids provided by man and animals, e.g. deforestation, overgrazing, etc. one of the most important influences of man and his associated agents on the flora is that of a carrier of diaspores of weeds and exotics, several of which have naturalized in the Indomontane ecosystem and have become obnoxious element in the flora (e.g. Broom—Sarothamnus scoparius Koch, Gorse—Ulex europaeus L., Ambrosia artemisifolia L., Erechties valerianaefolia DC., Kew Weed-Galinsoga parviflora Cav., Synedrella nodiflora Gaertn., Soliva anthemifolia R.Br. ex Less., Tagetes minuata L., Nicandra physaloides Gaertn., Eryngium foetidum L., Axonopus compressus Beauv., Purple Cudweed-Gnaphalium pennsylyanicum Willd., Erigeron canadensis L., Lagascea mollis Cav., Cat's Far-Hypochoeris redicata L., Silybum marianum Gaertn., etc.).

Some adventive weeds introduced from tropical, equatorial and South America, Europe, etc. have become pests and 'green cancers' cousing agricultural ecological and health hazards from time to time (e.g. Lantan camara var. aculeata Mold., Eichhornia crassipes Solms, Eupatorium odoratum L., E. adenophorum Spreng., Mile-a-minute climber—Mikania micrantha H.B.K., Thickhead Crassocephalum crepidioides S. Moore, Parthenium hyspterophorus L., Hyptis suaveolens Poit., Alternnthera pungens H.B.K., Argemone mexicane L., Xaanthium pungens Wallr., etc.).

Absence of Natural Enemy

The most important factor attributed to the spread of exotic weeds in Indomontane ecosystem is absence of natural enemies (e.g. fungi, birds, mammals and insects) as a result of which these plants have proliferated rapidly. In view of the seriousness of this growing problem, there is urgent need for an operational programme at the national level involving intensive survey of areas heavily infested with weeds, pests and neophytes in themontane ecosystem, followed by systematic ecophysiological studies and an effective eradication programme to be adopted on a large scale in the country.

VIII. MALARIA PROBLEM IN THE HILL DISTRICTS OF WEST BENGAL AND ASSAM

Incidence of malaria cases are increasing both in the plains and in the hilly areas in Eastern India. A total of 77,332 malaria cases have been reported from all over W. Bengal (since 1971 to 1975), of which 19357 (24.88 per cent) occurred in the hill districts. There has been a drastic change in the environment of the hilly regions due to rapid development during the last few years. Replacement, disappearance and new additions to the mosquito fauna has occurred as a result of change in the ecosystem. Malaria in hilly areas has created many problems, come of which are discussed in great detail in this study.

Vector Borne Disease

The transmission dynamics of a vector borne disease is influenced by a number of factors, of which the environmental factor plays a very important part. Other factors responsible for successful transmission of malaria are studied.

Five Anopheline speciews(A. stephensi, A. philippinesis, A. culicifacies, A. fluviatilis, A. minimus) were reported as ventors of malaria in the hill districts of W. Bengal and Assam a decade ago. Three new additions (A. annularis, A. jeyporiensis, A. balabacensis) have been made to the existing list.

Malaria patients suffering from P. vivax, P. fulcipagum and those having mixed infection have been reported from Jalpaiguri and Darjeeling Districts. Non-human primate malaria paarasites, (P. knowelsi, Pscynomolgi, P. inui, and P. coatneyi etc.) can be transmitted to human beings experimentally. Natural human infection with primate malaria have also been recorded, hence the danger of such infections in the mountainous zones with increasing, cultivation and development has also been discussed.

Among the several haemoglobinopathies, Hb-S. and Hb-E thalassaemia patients are resistant to infection with P. falciparum and P. yiyax infaction respectively. Hb-S haemoglobinopathy is reported in the aboriginal tribes, and Hb-E thalassaemia is common in Assamese and Bengali population. There is however no direct evidence to show that Hb-E and Hb-s genes have any protective role.

Malaria

Malaria cases due to P. falciparum infection in Assam and Meghalaya areas have been reported to develop resistance to 4-amibiquinolines (chloroquine), which has further complicated the problem. *A. philippinensis* was tested against D.D.T. in 2 units in Assam and hill districts of West Bengal need to be carried out. The above problems are dealt in the paper to be presented.

Due consideration should be given to the problems arising from the change in ecosystem in the hilly areas in connection with the long and continuous battle against malaria.

IX. PRELIMINARY REPORT ON A SURVEY OF AQUATIC ECOSYSTEM OF THE RIVER BEAS, WITH SPECIAL REFERENCE TO INSECTS

This study is preliminary report on recent studies of the aquatic insects communities of the river beas. The studies were undertaken to form an idea of the composition and character of the Beas River ecosystem before the full impact of the technological developments. As a part of the beas-sutlej link project, the river Beas will in future become converted into a vast lake of great depth and sluggish flow between pandoh and bhuntar. The present studies help in forming a preimpact picture of the ecosystem.

X. CHIRONOMIDAE (DIPTERA) OF RIVER BEAS ECOSYSTEM

General features of the family Cfhironomidae (Diptera) and their habit and habitat are discussed. Results of the collection tours undertaken in the beas river ecosystem before the construction of the pandoh dam are provided. The changes likely to occur as a result of the construction of dam are briefly indicated.

XI. ECONOMIC GROWTH AND ECOLOGICAL EQUILIBRIUM-A STUDY IN THE HIMALAYA

History has repeatedly shown that when ecological changes take place in the mountains, changes soon follow in plains. Still mountain ecosystems have been ignored so long in comparison to other natural areas. There is immediate need for the formulation of a policy for the prevention of ecological degradation in the Himalaya. However, it is equally essential that such a policy should not be detrimental to the economic growth and development of these regions.

The approach, that the stoppage of economic growth can bring about environmental conservation, is not only impracticable but in the long run self-defeating. In the Himalaya,infact, environmental conservation and economic growth can be made complimentary. While taking into consideration the socio-economic structure of the Himalayan regions, this study deals with the formulation of a developmental policy which besides bringing about economic growth of these areas would lend protection to their ecosystems.

XII. CHANGES IN THE SOIL-INSECTS COMMUNITY STRUCTURE UNDER THE IMPACT OF ROAD CONSTRUCTION AND DEFORESTATION IN THE NORTHWEST HIMALAYA

This study describes the changes that have taken place during the last twenty years in the soil-insects community structure in the upper reaches of the Kullu Valley and the Lahaul Valley, under the influence of deforestation and high altitude road construction. The kullu valley has been connected with Lahaul and Ladakh valleys for the last many centuries by a trade route which crossed numerous Himalayan passes over an altitude of 4,000 m above the m.s.l. this trade route permitted traffic only by pack-animals like mule and yak. After 1960, however, the trade route with minor alterations in alignment was converted into fair vegicular traffic.

Soil Ecology Along Roads

This road, a marvel of modern technology, is now considered as one of the highest mountain roads in the world. As a result of road construction in the Lahaul valley and the upper reaches of kullu valley, accompanied by deforestation of mountain slopes (particularly in the upper reaches of kullu valley), the soil ecology on either sides of the road has undergone tremendous changes. Consequent to these changes the soil–insects community structure has been profoundly altered. For instance, prior to the construction of road and the deforestation, Collembola constituted the predominant group of insects in the soil-insects fauna.

A study carried out during the years 1975 and 1976 has revealed that due to the erosion of soil in the deforested area, blasting of rocks and mountain sides and construction of retaining walls has considerably reduced the humus formation in the soil all along the road. Consequently collembola which depended for its existence on the rich humus conditions has shown a tendency of recession in its distribution. Another distribution trend noticed recently has been an almost complete obliterating of the dominant or indicator species of soil coleoptera, viz calosoma madeirae from the denuded area. The probably causes of all these changes are discussed in this study.

XIII. ENTOMOLOGY EXPLORATION IN LADAKH HIMALAYA

Ladakh Himalaya offer unique opportunities to observe the adaptations met with in insects in extremes of environmental conditions. The general area is so vast and devoid of vegetation that insect life is confined to pockets of favorable climatic conditions, thus offering opportunities to study their distributional patterns. Unfortunately, this area was a restricted area for a long time and not much scientific explorations have been done there. We have not much background information about the composition of the fauna and flora, except for some earlier explorations during the early part of the present century. Only recently the area has been opened up for tourist traffic and the activities of mankind will make an impact on the fauna and flora of the region.

It is therefore very important that many scientific explorations are made there as soon as possible so that we gather data of importance before the ecology of the area is changed.

Road Building

There have been activities of road building to a large extent by the army due to the military importance of the area. Yet I feel that not much change has taken place in the general ecology of the area, as the population pressure is absent. With this view in the mind I made a trip in 1937 to gather insects and study the ecology of the area first hand. My visit was rather of a preliminary nature, but this gave me an insight into the faunal composition of the area and what changes are likely to occur due to the activities of mankind, chiefly army personnel.

XIV. BIOECOLOGY AND DISTRIBUTION OF TERMITES IN THE HIMALAYAN REGION

The Himalayan mountain system as covered this study ranges from kumaon Hills (in uttar Pradesh) to Kashmir in the northwest Himalaya, also known as the "extra peninsular region", comprises a series of mountain ranges lying more or less parallel to each-other with valleys in-between. Of the four parallel zones, the distribution of termites is restricted to the first two zones, namely shiwalik zone and sub-Himalayan zone. Our knowledge of termites of this region has been greatly enriched by the contributions of Desneux (1904, 1906); in (1919) ; Chatterjee and Thakur (1963-1968); Gupta (1975); Sen Sarma et al. (1975) and Thakur (1973-75). The region is represented by 39 species and 17 genera.

All the major families are represented as follows:—(i) Kalotermitidae (two genera and four species); (ii) Hodotermitidae (one genus and two species); (iii) Stylotermitidae (one genus and four species) (iv) Rhinotermitidae (two genera and two species) and termitadae (eleven genera and twenty-seven species).The paper discusses in detail the bioecology, distribution and the economic role of termites in the region.It has been observed that with the rise in altitude, there occurs a product decline in the number of genera and species. This conforms to the known pattern of distribution of termites as a whole, as relative humidity and temperature, the two keys a biotic factors, mainly govern their distribution. This also suggests that the Himalayan region, particularly beyond 2000 meters is not the natural home of termites.

Termites

Termites, like any other life community, form an integral part of the ecosystem. It is therefore, of necessity that conclusion if any have to be drawn in full and by as an individual isolated group. Termites are perhaps the longest lived social insects and antedate man by millions of years. However, by his sheer ingenuity, man is rapidly acquiring a firm grip over this adaptive horizon. He has broadly manipulated the environment and has introduced far reaching changes in the sylvan biocoenosis.

The clearing of more and more of forests, putting larger area under cultivation, introducing monoculture concept of plantations, construction of dams, tunnels and blasting operations in the hilly areas for road constructions etc., all these have tended to set in a chain of interactions,which sometimes, may wipe out the form of life community from an area or forces changes in their habitats and habits. Changes in climate, subsoil conditions etc. Occurring mainly due to the activities of man may thus account for occasional sudden local increase in termite damage. The disturbance in the balance of nature by man improves the living conditions of his enemies and termites are no exceptions to the rule.

Bio-ecological Change

Quite often, the bioecological changes in life communities, coupled with the topographical changes of great magnitude, have changed the very characteristic of termite fauna of an area and today, termites are considered to be one of the greatest destroyers of human property. From forest and field scavengers, they have become pests of man's belongings, largely because of his careless constructions of buildings and his failure to take adequate measures against them. The reported damage to buildings in askot (kumaon hills) threatening the whole town ship is a classical example worthy of mention.

The study also discussed the beneficial role played by termites. Great many species of termites are subterranean in habits and are so very numerous in number, feeding exclusively on the dead vegetation, that they join hands with other agents such as micro fauna of the soil, relative humidity and temperature and accelerate the process of disintegration of all lifeless plant matter. They are therefore, an important factor in assisting the growth and continuos renewal of plant life.

The role-played by the species of *Archotormopsis* in disintegration the decaying wood of logs. In the himalaya, can well be appreciated in the highly difficult terrain where logging operations are not only uneconomical, but also full of risks. Therefore, it is in the interest of man that he learns to exploit the biosphere more judiciously, with the minimum possible interference in the natural balance and thus helping to a great extent in preserving and maintaining the natural panoramic beauty of the mountain system.

XV. OBSERVATION ON THE EFFECT OF HUMAN INTERFERENCE ON THE AGROMYZID POPULATION IN THE MOUNTAIN ECOSYSTEM

If any ecosystem, insects undoubtedly have a dominant position. Mountain ecosystem is no exception. Numerically insect dominance is unquestionable. Their wide range in time and space has been possible due to their extreme adaptability and intricate involvement in their particular ecosystem. Nature has put many demands on insects especially on insects in the higher altitudes. In the case of Fold Mountains like the Himalaya, their initial abodes were gradually lifted up from lower levels to higher altitudes. But insects have efficiently overcome and adapted to the changes posed by the process of nature. This we can see if we make a careful study of their ecology.

All the three recognizable component groups of mountain insect fauna, the original inhabitants remaining as such without any change, the original inhabitants remaining as such without any change, the original inhabitants that have undergone changes, and the colonist from the plains, all equally affected by multifarious human activities.

Agromyzinae Population

This study contains a detailed analysis of the present day agromyzinae population of the mountain ecosystem numbering 18 species. Most of the agromyzid are host specific and in many cases even topospecific. More than 60% of the agromyzid found in the mountain. Communities are introduced by human agency as a result of the introduction of their respective host plants. The paper deals in detail with the human introduced species of the genera melanagromyza hendel, tropicomyia spencer, ophiomyia brashnikov and japanagromyza sasakawa belonging to the sub-family agromyzinae and cromatomyia hardy belonging to the subfamily phytomyzinae. Also included in this study are some general observation on the agricultural practices responsible for large-scale distortion of insect population and some suggestion for future approach.

XVI. IMPACT OF TECHNOLOGY ON THE SOCIO-ECONOMIC FABRIC OF THE LAHAUL VALLEY AND ITS EFFLORESCENT EFFECTS ON THE KULLU VALLEY (HIMACHAL PRADESH)

The twin valley of. Lahaul and Spiti, constituting a district of Himachal Pradesh, are situated to the north of Pir Panjal range. To the north are the zaskar and Ladakh valleys while on the south kullu and chamba valleys bound the district. Of the two, the Lahaul valley is much larger in area and population and forms the basis of this study.

Till 1960 the valley was almost practically free from any influence of modern technology and scientific agricultural practices. The entire valley, through which meander the rivers Chandra and Bhaga and down and to the west the upper chenab river, practiced primitive agricultural methods primarily due to the inhospitable condition of climate and terrain. The main produce used to be buckwheat (local kathu), wheat, barley and some potatoes.

Medicinal Herbs

A medicinal herb (locally known as kuth) saussurea lappa, used to be grown as the main cash crop. The root of this herb was exported to Tibet and china for some unknown purposes. This crop used to bring in a lot of money to the valley but only to very few rich farmers who could afford to wait for three years for the maturing of the root-as it took three years to take the root export worthy. During all this period nothing else could be cultivated in the kuth growing fields. Other crops used to be hardly sufficient for a small population that lived much below the poverty level.

In the upper reaches of the Chandra and Bhaga valleys no forests grew. Lower down, however, dwarf junipers and tree junipers (juniperus communis) apart from few birches (batula utilis) were encountered. In the Upper Chenab part of the valley pinus excelsa forests on some slopes also grew. Near the villages and hamlets some willow trees (falix fragilis), reported to be introduced by the Moravian missionaries in the last century, were also grown.

The population remained under check due to the socio-religious institutions like polyandry and the institutions of chomos (budhist nuns). It is very significant to note that in fifty years (from 1901 to 1951) the population increased from 10,444 to 12,728 only. The literacy level was also extremely low.

Then came 1962, and under the geo-political pressures the valley witnessed a massive technological interference in the succeeding years. This interference consisted of construction of high altitude motorable road over difficult mountain passes connecting Lahaul with the kullu valley to the south and with the Ladakh region to the north. This road apart from providing outlets to the people also helped in opening up of the valley for further interference in the form of mechanized farming, reclamation of fresh area for cultivation, and introduction of high yielding varieties of cash crops in place of traditional kuth.

Technological Interference

This technological interference brought about a very impressive economic prosperity during the last fifteen years. The economic boom in its turn had a corroding effect on social institutions like polyandry resulting in a spurt in a population growth besides other factors. The efflorescent effects of this economic prosperity and abrupt growth of population are being felt in the adjoining valley of kullu. The causitive factors responsible for the changes in the socio-economic conditions and their impact on the kullu valley are discussed in detail in the paper.

XVII. IMPACT OF LAND USE CHANGE ON THE OVERALL ECOSYSTEM IN HILLS OF NORTHWESTERN INDIA, WITH SPECIAL REFERENCES TO HIMACHAL PRADESH

Himachal Pradesh, which came into being as a separate political entity on 15th April 1948, was one of the most backward areas in the country. After its creation, however, large-scale developmental plans were put into operations. These plans dealt with the construction of roads, expansion of horticulture, construction of multipurpose river valley projects, defenses along the borders and changes in grazing pattern. As a result of implementation of these plans tremendous changes took place, which have transformed the ecosystem of this mountainous area very drastically. These changes are discussed in details in the paper with the help of authentic data. The author further discusses the corrective measures, which have been taken under the centrally sponsored programmes. These measures included the establishment of wild life sanctuaries, afforestation and introduction of the New Forest policy known as the three-dimensional forest farming.

XVIII. THE NEED FOR PLANNED AND INTEGRATED DEVELOPMENT OF HILL REGIONS

The emergence of human settlements in hill regions is indicative of a close inter-relationship and forms an important link between the man and mountain, i.e., promotion of tourism, defence against enemies, identifications of resorts and to derive economic benefits from the mountain ecosystem.

These dormant and unifuntional towns in the hill regions are witnessing rapid growth of population with significant changes in socio-economic structure i.e. higher disposable incomes, increased leisure and mobility. The positive forces of development have brought along with them the dangers and fears of disrupting the present ecological balance. Tourism in India is largely dependent upon the variety of nature, variation in climate provided by the hills and is now recognized as one of the fastest growing industries and the tenth largest foreign exchange earner. Promotional based concept of developmental strategy utilising the sectoral approach of developing hill towns to ensure better standerds in mobility and residential accommodation need to be altered.

Integrated Approach

An integrated physical planning approach with a emphasis on conservation and systematic analysis of terrain, forestry and economic analysis of providing tourist infrastructure should be adopted. Selection of new centres for promoting tourism to improve the socio-economic life styles of hill populations should also from an important component in the developing a closer and meaningful interaction between man and mountain ecosystem.

7

The Ecology of Kumaon: A Paradise in Peril

The Himalayas are one of the youngest mountain systems in the world and, in fact, younger than many mountain systems even in India. For instance, the Aravalli, Nilgiris and Vindhya ranges in the country are estimated to be 1500 to 2500 million years old whereas the Himalayas are only 40 million years old. Geologists believe that the Himalayas originated in a vast sea basin adjacent to a land mass. There is definite geological evidence that vast stretches of the Himalayas were originally covered, several million years ago, by what was known as the Tethys sea. Even recently, rocks of marine origin of the Cretaceous age (60 to 80 million years ago) have been discovered under the snows of Mount Everest and other great mountains. Rivers deposited vast quantities sediment into this sea, which started subsiding at the bottom.

Magma

After perhaps a few million years, the depth of the basin brought it into contact with (Magma) the hot interior of the earth. This resulted initially in a compaction of the lower layers. The Himalayas emerged from under what was originally the sea ass by a series of 'uplift' phases. These series of geological phenomena ultimately resulted in the sea waters draining away, and the rising mass of sediments rose to form a mountain chain. The different parts of the mountain chain were thus formed. According to geologists, the Himalayas are still rising and, therefore, are inherently unstable.

In case the equilibrium of the mountains is disturbed due to any cause, stresses build up within the apparently solid and strong rocks until a critical point is reached when the stresses are suddenly released resulting in displacement of rocks. Many earthquakes are caused in this way in the Himalayan ranges.

Youthful Himalayas

Most people think of the Himalayas as ancient and strong, and in fact as a mighty protector of the country in the north. In reality, the Himalayan mountains are geodynamically sensitive, youthful as compared to other mountain systems of the world, and not particularly strong in a geological sense. The Himalayas largely consist of weak, deformed rocks which are severely stressed and easily susceptible to the onslaught of heavy rains, and the vibrations and disturbances caused by the movement of heavy traffic and human interference.

The ecological balance of the Himalayas is very delicate, and any type of developmental programme has to ensure minimum adverse impact on it. These mountains therefore deserve great care when considering excavations, major road-building exercises and heavy construction. We should also avoid creating further environmental hazards resulting from the removal of protective forest and vegetation cover. As we shall see in the course of our discussion in this and other chapters, the nature of the Himalayan mountain system has effects on all aspects of development in the hill areas. Himalayas are the mainstay of people of hill areas.

General Characteristics of Climate in Kumaon

The Himalayas presents a wide variety of atmospheric conditions, which vary according to the direction of mountain valleys, alignment of ridges, and relative differences in the, elevation and altitude of the mountain ranges. In fact, remarkably wide differences are noticed in the Himalayan region even over short distances. There are broadly two types of climate in the entire mountainous region. These are designated by climatologists as the moderate climate zone, and the alpine climate zone. 'Moderate climate' prevails in heights up to 3000 metres. The 'alpine zone' covers areas above 3000 metres where the only inhabited parts are the alpine valleys where large tracts are covered with snow during winters; the alpine hills represent extreme conditions, with a perpetual snow cover.

Climate Differences

Even in the moderate climate zone, there are variations in the climate depending on whether we are talking of the hills or the valleys. Both pressure and temperature are affected by the altitude, both decreasing as the altitude increases. Rainfall in Kumaon also manifests great variety and is dependent to a large extent on relief features, mainly the nature and alignment of the mountain ranges, and not only on the altitude. To take, one example, the average annual rainfall in Ranikhet is less than half that of Nainital though the altitude of the two places does not differ to any great extent. The year is generally divided into three main seasons: the cold season from November to February, the warm/hot season from March to mid-June, and the rainy season from mid-June to October.

Earthquakes, Erosion and Landslides

The Himalayas have had a few major earthquakes in the last one hundred years, and experts have identified several potential earthquake-prone areas known as 'seismic gaps' where the strain is building up progressively. The

Kumaon region itself has had no major earthquake in this century, and the nearest major earthquake in recent years had its epicentre in Agora in Uttarkashi in the Garhwal Himalaya. But the multiplicity of geological faults and the underlying geological instability referred to by geologists as 'tectonic ferment') in these mountains, hold the ever present threat of danger. In order to reduce the impact of future earthquakes, some experts have suggested the preparation of hazard-zoning maps in which the areas of potential hazard are identified.

In addition to potential earthquake-prone areas, Valdiya points out that it is important to identify areas and belts prone to rockfalls, landslides and debris flows on the basis of geological features, rainfall data, past history and current signs of slope instability. This could help in relief and preventive measures to be taken expeditiously at times of crisis, and also to take into account the geological sensitivity of certain areas in the construction of new roads and buildings. As far as earthquakes are concerned, it is perhaps not possible to do much more about them than to provide relief quickly when they do occur. But as far as landslides are concerned, it is certainly possible to take some preventive action.

Slope Instability

For instance, every new road or large construction in the hill areas should be examined from the point of view of slope instability and other geological features before going ahead with it. The construction and planning of all buildings must also take into consideration the forces likely to be generated by earthquakes. While hazard-zone maps may help communities to prepare plans to meet potential earthquakes, the more important daily threat faced in these mountains is that of landslides.

Valdiya has pointed out how the geological problems of the unstable mountains have resulted in the gradual but persistent 'uplift' of the terrain, which has significantly quickened the rate of erosion. Precise figures are not given but the present rate of erosion is many times higher than in the immediate geological past, and it is gradual increasing. Such higher erosion has the effect-of increasing the instability of the hill slopes.

In addition to this accelerated erosion owing to geological, reasons, there is the erosion of top soils because of deforestation and reduction of vegetal cover. The accelerated erosion rate has the consequence of voluminous sediment production, filling valleys and reservoirs. The figures are staggering. In one catchment area alone in these hills, every 100 square kilometres is reported to produce nearly 23 hectare metres of sediments annually. The sediments take away the top soil rich in nutrients. The process of erosion also weakens the mountains, which become more susceptible to landslides than before.

Haphazard Construction

Inadequately planned road construction, haphazard building constructions and continuing deforestation aggravate the threat of landslides. Even during the preceding two years 1998 and 1999, there have been many big landslides in the Kumaon hills. A landslide in Nainital town recently destroyed part of a road on the hillside above the less populated (and undisputedly more beautiful) side of the lake. Another major landslide at night near the border of Pithoragarh district ended in major tragedy, killing more than 200 pilgrims on the route to Kailash Mansarovar. These are warnings and pointers to a grim and dangerous future if timely action is not taken. It is, in fact, clear to anyone familiar with the Kumaon area that the construction of new buildings does not usually take into consideration environment-related concerns which are genuine and urgent.

For instance, under the UP Roadside Land Control Act, any construction within 100 feet of the centre of a PWD road is totally forbidden. Yet one has only to drive a short distance on any of the main roads in Kumaon to see numerous residential buildings and shops very much within this distance from the road, some of them old, but many in the process of coming up without any visible effort on the part of any public authority to stop them. It is difficult to believe that the concerned officials are not aware of these constructions, which are not only illegal but add to the potential instability of the hill slopes. This is truly a man-made and, therefore, avoidable environmental hazard. There are many such instances of unplanned development which one comes across on a daily basis in the Kumaon hills. Many 'resorts' and residential complexes which are totally unplanned and, in some cases, haphazard are coming up Outside the municipalities and cantonment areas.

No approval of any authority is required for any construction outside the municipal or cantonment areas, and it is not dear as to what extent considerations of slope stability and related environmental aspects have been taken into account in planning these complexes. The authors are personally aware of one residential development near Ranikhet where regular water supply is not available, i.e. there is no water connection from any rural or other scheme. Many of the houses have been bought by persons from Delhi and elsewhere. Whenever the owners or their guests stay in these houses, which is usually for a week or so during each visit, water for domestic requirements has to be brought in tankers from about twenty kilometres away.

As far as new construction activities are concerned, it does appear necessary to have a 'planning authority' to approve all new constructions in the hill areas so that considerations of slope stability, landslide hazards, water supply, etc. are kept in view. Architects or certified civil engineers familiar with hill construction may be required to certify in the case of each new building that its location, plan and other characteristics satisfy the considerations of slope stability, etc. and that sanitation and water requirements have been taken into account. This may help to restrict haphazard construction activity in the hill areas. If construction goes on as it is now, it can do a great harm to the ecology.

Water

Water has a crucial role in all domestic, agricultural industrial activities. The geological instability and ecological, sensitivity of the Himalayas as well as the fact of its being the source area of all major rivers of north India gives add significance to the availability and management of its water resources. Prima facie, there ought to be plenty of water in the Himalayan region for irrigation, drinking and other daily requirements. In fact, more than a billion people in four countries-India, Pakistan, Tibet and Bangladesh-depend in one way or another upon water from the Himalayan rivers. But the inhabited Himalayan areas suffer from a chronic shortage of water, to which theme we will be returning repeatedly in the course of our discussion.

Water supply schemes in many villages have become inoperative because of poor maintenance and other factors such as reduced flows in the streams and the springs. A study conducted by Tiwari in the Upper Kosi catchment area, which includes Almora town, showed that only about one-third of the recommended requirement of domestic water supply (120 litres/per person/ per day for urban areas and 90 litres for rural areas) was actually available. The situation is broadly the same throughout the Kumaon hills. Valdiya says: "this 'too little too much water' syndrome, is characteristic of a desert country." Another writer says that "the perennial problem posed by water in the region lies in-this duality-surfeit-and scarcity".

Sources of Water

The main sources of water supply in the hills are the mountain rivers, which in turn derive their water partly from the melting of snows and glaciers from the monsoon rains and the hill springs which are analogous to groundwater in the plains. There is an abundance of flow in the Himalayan rivers-which consist of three major river systems, viz. the Brahmaputra, the Ganga and the Indus. It is estimated that these three systems have an average annual flow of 1200 billion cubic metres.

Ganga System

For the Kumaon region, the Ganga system is relevant, some of whose major tributaries such as the Kosi and the Ramganga flow through it. The Ganga system alone has an estimated annual flow of 460 billion cubic metres, of which nearly-half can be used for irrigation. This compares very favourably with the irrigation potential of the Brahmaputra basin (a mere 3 per cent) and the Indus basin (about 25 per cent). Though some of its major tributaries flow through the Kumaon hills, the main source of the Ganga is the Gangotri glacier in Uttarkashi district in the Garhwal region of the Himalayas, where it is called the Bhagirathi; the other source of the Ganga is the Satopanth glacier near Badrinath, where it is known as the Alaknanda.

The two rivers merge at Devprayag, also in Garhwal, to form the Ganga. It then flows through Uttar Pradesh, Bihar and West Bengal. During the course of its 2500 kilometre journey, the Ganga bifurcates into the Padma (Bangladesh) and the Hooghly in its lower reaches, flowing into the Bay of Bengal. The Ganga basin catchment area, which includes its many tributaries, is nearly 11 lakh square kilometres, of which nearly 80 per cent is in India. The Ganga basin itself covers more than one-fourth of India.

Peaking Rivers

In Kumaon, the rise in snow-fed rivers occurs from March, with peak flows in June and July, whereas the rise in rivers originating below the snow line commences from early July, peaking in August and September. Generally, all rivers are in spate during the monsoon-season, and 70 per cent of the total river discharge is noticed in July, August and September. Because of the nature of the topography and the force of the water-flow down the hills, the potential of generating hydroelectric power is considerable in such territory, and the Kumaon hills are no exception. For the Ganga basin, for example, a power potential of nearly 11500 MW (at 60 per cent load factor) was estimated in 1981 though the actual realization of this potential is still quite small.

Hydroelectric Power Projects

Of course, implementation of more hydroelectric power projects could inevitably bring in its wake various environmental problems These potential projects are located both in the Kumaon and the Garhwal regions. Some of these projects are underimplementation, but the untapped potential is still considerable. It must also be mentioned here that the irrigation potential is largely realizable only in the plains, though additional power can be utilized anywhere in the grid, including the hill region. However, the real need of the people of the hill region is water because the power requirements of the region are quite small, as industrialization in this area has so far been very limited. From another point of view, whenever any new power project starts it represents a source of income-generation for the new hill state.

Springs

Springs and mountain streams are often the only available water source in the hill areas for agriculture, horticulture as well as domestic uses. While precipitation in the form of snow-accumulation on the mountain slopes forms the principal source of water in the higher altitudes, rain water is the main water source in the lower mountain ranges. Rain water adds to the surface flow of the rivers and streams, and also percolates and 'infiltrates' under the ground to emerge as springs. The nature of the soil structure is very important. Infiltration is greatest in the relatively rough, permeable and more gently sloping terrain.

Almost all perennial springs and streams are located generally on the 'fault-lines' or 'fracture' of hard impermeable rocks. While there is a vast quantity of water flowing in the Himalayan rivers, most of the water frequently runs down the hills, only to cause floods in the plains. The scarcity of water in the hill region is a real one as there is an enormous difference between water flowing during the dry and the rainy seasons. One expert says that, in streams fed by springs in the lesser Himalayas, the difference may sometimes be more than 30000 times! Even modest estimates put the difference as more than 1000 times in Most cases.

Spring Replenishment

A spring in the hills is replenished continuously by rainwater seeping through the soil as well as seepage from river water in the surrounding areas. This is known as 'recharge' of the spring water and is somewhat similar to the recharge of the underground water which provides water for tube-wells and pump-sets in the plains. For a spring to have adequate flows of water, the recharge must be adequate and continuous. Springs where the water is available round the year for utilization are known as 'perennial springs'. Many springs dry up during the course of utilization. In fact, more than 85 per cent of rainwater in most areas flows down the mountain slopes as surface-runoff, and less than 15 per cent is available to recharge the Himalayan springs on which large numbers of hill people depend.

Reduction of Water Absorption

Deforestation and poor agricultural practices have resulted over the years in the reduction of the water-absorption and retentive capacity of the land. In a field study of 49 springs done by Tiwari, it was found that 18 had completely dried up during the last 10-15 years due to these reasons. Villagers have also reported that the flow in the, remaining springs has reduced considerably during the last two decades. Greater absorption by the soil will occur if the vegetation cover increases, i.e. greater afforestation will reduce the surface run-off and make available more water to recharge the Himalayan springs. Augmenting the water flows in the-springs by the development of spring sanctuaries has been suggested by one expert on the subject. The idea is to enclose the catchment areas of different springs with appropriate fencing to prevent access of grazing animals as well as fodder and timber gatherers. As part of the augmentation plan, trenches ands pits will be made in order to promote infiltration of more water which can recharge the spring water at a greater rate than before.

Rainwater Harvesting

'Rainwater harvesting' is a term used to denote the systematic and regular collection of rainwater basically for domestic use. In suitable cases, underground reservoirs and dugouts can be used for storing rainwater. Of course, rainwater harvesting can be done in the plains areas as well, but it appears to have greater relevance in the hill areas. Many households in Kumaon have already started rainwater harvesting from their roofs and the movement is gathering momentum. Some non-governmental organizations are also engaged in the construction of rainwater storage tanks in the hill areas.

Rainwater can be an important source of household water supply for uses other than drinking and cooking. It cannot normally be utilized for drinking purposes as it lacks many essential ingredients necessary to make it suitable for human consumption, but it can be used for bathing, washing, gardening, etc. An ideal way of using rainwater is to gradually utilise it over the dry months from the storage reservoir at a rate that will ensure that the stored water lasts till the inflows in the subsequent rainy season. The collection and storing of the rainwater, if done on a large scale in the hill areas, also means less 'surface runoff' and, consequently, less topsoil erosion.

Glaciers

Glaciers are formed from the accumulated snow which has survived melting during the summers. When the snowfall during the winter is greater than the quantity of snow that melts during the following summer, a glacier can be formed. One writer on the Himalayas mentions that there is no specific word in 'Indian languages for 'glacier', though the hill people have local names such as bamak, gal, etc. A large number of glaciers may be found above the permanent snow line, i.e. 17000 feet above sea level. The formation of a glacier depends on area-specific factors such as the height of ridges, slope-orientation and the quantum of precipitation.

For example, on account of differences in the solar radiation, south-facing slopes have usually less ice-cover. Nearly forty years ago, it was estimated that nearly 17 per cent of the Himalayas, i.e. about 33000 square kilometres, were under lacier ice. There are an estimated 15000 or so glaciers in "the Himalayas, some of them being 1 to 5 kilometres in length. Some glaciers such as the Siachen, are as long as 72 kilometres, and the Gangotri in Garhwal is 26 kilometres long.

The glaciers in the Kumaon region are smaller. In Kumaon, the Milam glacier is 19 kilometres in length and 3 kilometres in width, and the Pindari glacier is 3.5

kilometres long and 500 metres wide. As already mentioned, the Himalayan glaciers are also a very significant source of fresh water, a substantial part of which goes into feeding the-rivers near their Himalayan sources. An idea of the water potential of a glacier can be had from the fact that Gangotri glacier has an estimated volume of ice which is more than twice the total capacity created by the Bhakra dam. In western Himalayas, of which the Kumaon region is a part, the upper reaches of the major rivers do not get much monsoon precipitation and the glaciers contribute substantially to the river water.

Melting Glaciers

The melting of the glaciers contributes more than half of the water flow in many of the Himalayan rivers. But the melt water is often wasted, as least as far as the hill region is concerned, as the melting reaches its peak in the late summer just before the rainy season which follows it very closely. In fact, while glacial melting can supplement a weak monsoon, it can aggravate a flood situation where the monsoon is strong. The melt water, while flowing down the hill slopes, does augment the flow in the Himalayan rivers as well as the springs.

A substantial part of it flows down the slopes as 'surface runoff' and is wasted as far as the hill areas are concerned. Perhaps because of 'global warming', many glaciers are receding and a vast frozen asset is reducing with each passing year. Observations by some experts support this trend, e.g. the Pindari has shrunk significantly during the last one hundred years while the Milam glacier has 'retreated' by nearly 1.3 kilometres during the same period. On the other hand, some recent studies appear to indicate that some glaciers have begun to increase in size again, which is indeed a hopeful sign.

Storage of Melting Water

As the peak period of melting coincides generally with the rainy season, some methods will have to be found to store this melt water by suitable dams at high altitudes, thus enabling the use of this water at the time of need. One expert has suggested glaciological and hydrological investigations for the preparation of suitable water-resource development programmes. It has also been suggested that such dams, as they will be located in very sparsely populated higher Himalayan altitudes, may not cause the environmental problems that similar projects can create in the populated lesser Himalayan areas. As can be readily imagined, glaciers are a great attraction for those tourists who are interested in trekking in the mountains. They thus form an important part of what is generally called 'adventure tourism'.

Biodiversity

Any discussion of the environment will not be complete without referring to biological diversity, usually termed 'biodiversity', which denotes the multiplicity of life forms which sustain our planet. Biodiversity is the "property of living systems of being distinct, that is, different, unlike." It refers to the variety and variability among living organisms and their surroundings. Three categories of biodiversity, viz. ecosystems, species and genes, are usually distinguished. Ecosystem biodiversity refers to the variety of habitats in the biosphere; e.g. a landscape which has croplands, grasslands and woodlands has more diversity than another With just fragmented woodlands. Species biodiversity concerns the different species prevalent in a specific geographical area. Genetic diversity concerns the genetic varieties of plants, animals and microorganisms.

Conservation of Wild Species

Conservation of wild species is therefore an important part of the biodiversity conservation programmes. A large number of plant species have productive potential in the pharmaceutical industry. Many have already been used in drug formulations, and unless this biodiversity is sustained by careful and planned exploitation, posterity may lose part of an invaluable resource. Apart from innumerable wild edibles (334 varieties in central. Himalayas alone are listed by experts) such as roots, tubers, flowers, fruits, seeds, etc., which are nutritionally important, a large number of fodder species exist in the Himalayas. Wood, which is still a major domestic fuel,-is dependent on a variety of forest resources which are under threat of reduction owing to several (human) causes. The Himalayan region supports over 80 species in the timber category alone.

Apart from the above, culture collections of a number of Himalayan species of micro-organisms (bacteria, viruses, algae, fungi, etc.) are used in metal and coal mining operations, cleansing oil spills, controlling pests, etc. These are only a few random examples to illustrate the significance of biodiversity.

Change in Biodiversity

A good example of a major change in biodiversity in the Himalayas, especially important in Kumaon, is the shift from banj (oak) to chir pine forest as the dominant species in many areas. Banj has innumerable uses for the hill people. It is a very good charcoal and firewood source, though it is slow-growing. It also provides abundant fodder, being a broad-leaved species. The leaves also generate a thick humus, is an excellent fertiliser. Its very capability to meet multiple—needs has led to its over-extraction for firewood and charcoal, lopping for use as fodder and removal of the humus from the forest.

This over-exploitation combined with its slow growth results in poor regeneration and the forest ground being rendered bare. The importance of regeneration of the oak forest is not often realised as the failure of regeneration would not come to notice when the old trees are still there in the forest. It is therefore essential that the programme of dissemination of young oak seedlings is undertaken on a regular basis to ensure continuity of the forest in the long run. In some areas, oak forests are allowed to make way for apple growing and, in many others, to chir pine. The area thus becomes a favourable habitat for chir pine, which grows faster and has some commercial value for the forest department while not being of much use to the local people for fuel, fodder, etc. The shift from banj has further consequences for other flora and fauna associated with banj, and the whole forest composition changes gradually. Problems of soil erosion, water scarcity and loss of biodiversity invariably result from the loss or reduction of the oak forests.

Growth of Uncontrolled Tourism

Biodiversity also faces another threat from the growth of uncontrolled tourism. Many builders are buying land and building tourist complexes which do not often take into account the capacity of the available infrastructure to handle the needs of this extra influx of population. Visiting tourists are known to throw plastic bags and other non-biodegradable material which, in due course, is covered by the autumn leaves. Ultimately, this leads to the plastic material preventing the rainwater from seeping under the soil and augmenting the springs. Plastic accumulation can also prove destructive to the green life on the mountain slopes. The area between 3000 to 5000 metres above sea level, while it covers less than 5 per cent of central Himalayas, is very rich in species biodiversity but tourism and overgrazing have threatened the biodiversity of these meadows. Ajay Rawat has suggested that the extraction of medicinal herbs and plants should be banned in this area.

Values of Biodiversity

Biodiversity has a number of values for society, both direct and indirect. The direct values or benefits are obviously related to consumption, but the indirect benefits are more related to existence, keeping the options for posterity, and perhaps even survival of the human race on this planet. The loss of biodiversity is, to some extent, a worldwide phenomenon, but the problem is generally agreed to be most acute in the tropics, especially in the Himalayan region. Biosphere reserves, national parks and wildlife sanctuaries are invaluable in conserving important elements of biodiversity in the Himalayas. The relative importance of the Himalayas is seen by the fact that 6 out of 14 biosphere reserves are in the Himalayan region. It is essential that the importance of the maintenance of biodiversity in the Himalayas is recognized and accepted as an integral part of development planning. The recent 'biodiversity policy statement' of the Government of India, recognizes this importance.

Forests and Forestry

Vegetation and forest cover are of vital significance for the economy and ecology of the hills. They absorb water under the land surface, which in turn feeds the groundwater and especially the life-giving springs. They also bind the top-soils through the root systems and stabilize the atmospheric support in regard to moisture and carbon dioxide content, apart from providing energy to productive economic activity, especially agriculture. Forest cover in the hill areas thus performs several vital functions. It protects the slopes not only from soil erosion but also from landslides by binding the soil. It also improves precipitation by controlling the surface flows of water down the hill slopes. This helps to augment water in the springs and the streams.

The mountains and forests have always been intimately associated with one another. While 11.5 per cent of the geographic area of Uttar Pradesh (including the hill districts) is under forest cover, the hill districts alone have 44 per cent of their area under forests, according to official statistics.

Data on 95 out of 98 hill districts in the country shows that 30 districts have more than two-thirds area under forests while the remaining 65 districts have less than two-thirds. All the hill districts in UP fall in the latter category. The figures for Almora, Nainital and Pithoragarh districts of Kumaon, are 47 per cent, 53 per cent, and 39 per cent, respectively (including the newly carved districts of Bageshwar, Udham Singh Nagar and Champawat, which were originally parts of these three districts).

National Forest Policy, 1988

The National Forest Policy, 1988, emphasises the protective function of forests in maintaining ecological balance and environmental stability. It aims to have one-third of the area of the country under forest cover. For the hill areas, a higher figure of two thirds of the area is proposed, mainly "in order to prevent erosion and land degradation and to ensure the stability of the fragile ecosystems". On the basis of this yardstick, the forest Cover in e Kumaon hills is quite inadequate. In fact, many observers of the Himalayan scene believe that even the official figures of afforestation cited above are an overstatement of the field realities.

An unofficial expert study based on satellite imagery interpretation has estimated that only 28.7 per cent of central Himalayas is afforested and only 4.4 per cent has forests with a crown density of 60 per cent. These are much lower than the official figures. Forestry practices and policies have been widely criticised for the present unsatisfactory forestry situation in the hill areas. The afforestation and plantation plans have not kept pace with the widespread felling, both by the organised sector for commercial exploitation and by the local peasants for their energy and other needs. It has been pointed out that a tree takes minutes to fell and bring down, but more than fifty year to grow to its full size. S.L. Shah has estimated that, if the present rates of exploitation of the forests continue, natural regeneration will cease by 2031 AD. Similarly, grazing Pastures will be exhausted by 2041 AD.

The forestry policies were formulated several decades ago under the assumption of limitless forests meeting the needs of a limited population. This no longer holds good. Now the number of people per hectare of cultivated land in Kumaon, is nearly three times that of the plains of Uttar Pradesh. The population densities (which are much lower for the hills) do not give a clear picture as the cultivated and cultivable land in the hills is a much smaller proportion of the total area as compared to the plains.

Commercial Considerations

The main criticism frequently made of the forestry policies during the period of British rule and later is that they gave priority to commercial considerations over the needs of the local population. The forests were substantially used for timber needs of the industry; sal, teak, and deodhar were used for railway sleepers, and broad-leaf varieties such as oak (which were useful for local people in various ways, but not so suitable for commercial exploitation) gave way to conifers such as the pine, which were commercially more worthwhile.

This emphasis seems to have been continued in the forestry policies in the post-independence period as well. The most striking example of this slant was the encouragement given to pine forests and the reduced emphasis on broad-leaved varieties such as the oak. The consequences of rapid deforestation are many. The local populations are facing a firewood crisis, the nutritive top soils are getting depleted, the mountain springs are getting drier and the quickened run-off of the water down the slopes and the mountain rivers leads to floods lower down in the plains. The consequences for agriculture and sectors have been discussed in an earlier chapter.

Forest Fires

In recent years, forest fires have caused considerable damage to the eco-system of the hill areas. Forest fires have several adverse consequences. The water springs are not recharged adequately and they start drying up. Forest growth is retarded, the soil nutrients are adversely affected and erosion is given a further undesirable impetus.

Forest fires were unknown in the past, but there is no doubt that they have increased in recent years both in frequency and intensity. The villagers are usually blamed for most forest fires because they traditionally burn pine needles and pastures after summer to encourage the growth of fresh grass during the rains. This is also done by the villagers to make the grazing safer for the cattle on the slopes by reducing slippery litter. It has been recently reported by many observers of the hill scene that the villagers have developed a hostile relationship with the Forest Department, and do not often cooperate with the officials in controlling forest fires.

The hostility arises because the villagers perceive the department as denying them any rights in the forests, which traditionally provided them with fuel and fodder for centuries. While villagers have to be prevented from burning of forests for grass, it is also important to ensure that the 'timber and resin mafia' (to use Ajay Rawat's term) do not set fire to the forests for their advantage and put the blame on the villagers.

Destruction of the Himalayan Ecosystem

A lot has been written and talked about the gradual destruction of the Himalayan ecosystem and its long-term implications for the country. In the earlier parts of this chapter, we have tried to highlight some of the major ecological factors involved in the process. While we have concentrated on the Kumaon region, most of the problems are common to the entire Himalayan region. We shall now try to summarise these factors and examine briefly what can be done to arrest, and perhaps reverse, some of the adverse trends that have developed over the years.

Both natural and man-made factors have influenced the ecological system of the hills. Natural erosion is a major factor in the Himalayas. The precipitation levels caused by the monsoon (July-September) are without parallel in the world and the effects of heavy rains and water-flows down the mountain slopes of the Himalayas cause considerable natural erosion. The steep terrain, the instability of the young Himalayas and the heavy water run-off down the hills combine to cause tremendous erosion of the slopes.

Human Interference

One expert estimates that "if heavy rains double the water-flow, scouring capacity is increased four times, carrying capacity by thirty-two times, and the size of particle carried by sixty four times". The natural process is augmented by human interference in the ecosystem. Increased population pressure, excessive exploitation of forest resources, destruction of the mountain ecosystems by terrace cultivation practices, rearing of excessive livestock, certain types of road and building construction, eco-destructive practices of tourists-all of these contribute to this process.

The condition and state of the forest cover in the hills have major effects on the capacity of the forest to absorb and retain water, and also on soil quality. A good forest absorbs more water and ensures a lower water-flow down the slopes. When the monsoon is long and intensive, as in the Himalayan region, the condition of the forest assumes major importance. The speed of the water flows along the slopes also increases when the forest cover is reduced. This in turn increases the scope for landslides in the hill areas.

Steps Towards a Solution

Hans Christoph Rieger has pointed out specifically in the, Himalayan context that "the existence of a problem do not necessarily guarantee the existence of its solution". Many ecologists fear that the impending destruction of ecosystem of the Himalayas has by now become irreversible. However, we owe it to ourselves and posterity that even at this stage, we make all efforts to arrest it.

In reaching towards a solution of the Himalayan ecological problems, we must recognise the policy limitations clearly. We have to identify those variables that can be influenced by the government's policy measures and decide on suitable policies to influence them in the right direction. For instance, while we cannot affect the monsoon in any way, we can mitigate to some extent the intensity of the water run-off by structures such as check dams, river bank consolidation, etc.

Population Control Measures

Similarly, population control measures may slow down future population pressures in the hill areas and elsewhere. Human behaviour in the areas of forestry, construction, irrigation, cultivation and terracing practices can be suitably influenced by various policy measures in order to mitigate adverse environmental effects. Alternative fuel for domestic uses could reduce human incursions on the forest for firewood. More rapid afforestation is another long-term measure that is definitely a feasible policy approach to counteract the deforestation that has gone on for the last several decades.

Listing out some of the possible policy approaches, as we have done, brings out clearly the monumental nature of the problem. The problem is also substantially social, as evidenced by the behaviour of the hill population in all environment-related matters. An economist would say that this is a typical example of the difference between the individual and social points of view. From an individual's point of view, the felling of a single tree does no great harm but benefits him and his family immensely in the short run. But if each of a thousand individuals cuts a tree, it results in the destruction of a forest. What can be called future-ignoring behaviour manifests itself in the cutting or lopping of trees in the public domain. The present generation is taking a very short view of what is lies ahead. It is concerned with it own needs. What the future generations require is on the other side of the horizon.

Ignoring Social Costs

A similar case of ignoring social costs is seen in the decimation of public-owned forests. After all, if A does not fell (illegally, of course) a particular common-owned tree, perhaps B or C will do it. So it is privately rational for the first man (A) to cut and utilise the tree himself even though it imposes costs on society as a whole. The persistence of indiscriminate grazing in a publicly owned forest which needs a regeneration period is another such instance. If A does not graze his cattle, B or C would be likely to do so before he does so. It therefore makes (private) sense for A to grab the grazing opportunity before the others do so. But it is undesirable socially.

Therefore, what is required to resolve 'this dilemma of collective action' is either mutual agreement or coercion, or a little of both. This dilemma is best explained by the concept of the 'isolation paradox' first popularised in economic analysis by Amartya Sen long ago in one of his early books. This visualizes a situation where A may agree to a certain common policy restriction or measure provided B, C and all the others agree to it, but A individually will not agree to follow it himself. In the forest case, this would mean that A will agree not to cut trees in the common forest provided he is reasonably sure that B, C, etc. can be prevented by law or common agreement from doing so. This obviously would provide adequate justification for community action and active intervention by the government in conserving the forest wealth.

The role of public policy and effective intervention by the state and the central governments are therefore evidently crucial in the task of arresting the degeneration of the Himalayan ecosystem. The task is both figuratively and literally Himalayan in its dimension and scope. There is need for public awareness of the extent to which human interference has affected the Himalayan environment, and how this will affect not only the hill areas but the entire country for generations to come. We must take action before it is too late. As one writer on the Himalayas has observed graphically-"there is only one Himalaya to lose".

8

Rural and Urban Settlements

People living in the remote valleys of the Himalaya and in high glaciated regions and Alpine pastures are the people of nature. They and their ancestors have been living in the thick of nature for centuries. It is but natural that they show an uncanny skill in choosing correct sites for their settlements. The people from the plains who may have to choose sites for constructing new buildings for the Government may make mistakes in their choices. It is very important to have a correct appraisal of the environment to choose correct sites for new settlements and buildings. Otherwise it leads to disaster.

Mistakes in Locating Buildings

An example is given for the wrong judgement in just such a case. It is of a rest house near the Pindari glacier at a wrong place. The last rest house on the way to this glacier is Phurkia, about eight kilometres away from it. The Phurkia rest house is situated among huge boulders, resting precariously on the slope. The yawning gorge of the Pindar river is right below the rest house. It seems that the building may topple over any moment into the gorge. But it has not done so, because it is located on firm rocks. Further up, the slopes are made of loose morainic material.

From Phurkia the view is also not clear, as here the gorge is narrow. To a casual visitor a broad terrace about two kilometres away seems to be a much better site. It would have given a much better view, and would have been nearer to the Pindari glacier. The Public Works Department officers also decided to have a new rest house on this terrace. It was built, but an avalanche destroyed it after only one year. The bungalow was literally shot down into the gorge below. This showed that this site was an avalanche route. Also the area was made up of loose morainic material. When another bunglow was planned, the same site was chosen. Who knows when disaster would strike again.

A similar mistake has been made in siting Kalpa, the new headquarters of the newly constituted district of Kinnaur Kalpa has been located a couple of kilometres from Chini, the old Kinnaura village. It is situated on a gentle slope. But the accumulation of snow at this great height is natural, and in spring and early summer the snow accumulations slide downwards. Damage has already been caused to some buildings. Chini is, however, situated on a steep knife-edged spur away from the danger of avalanches.

Moreover, the location of Kalpa, high up above the, Sutlej gorge along which the road runs is also not correct. It is too high and makes transportation difficult. There is already a proposal to change the capital from Kalpa to Peu, five kilometres below and nearer the main road below. This situation perhaps is much better in all respects than Kalpa.

The rest house at Tola on the route to Milam was I also built so that it was right in way of valley winds. It was simply blown away. A very good site existed only a kilometre away behind a spur.

Choice of Village Sites

Choice of village sites is controlled by many factors, physical as well as cultural. Among physical factors the greatest control is exercised by land-forms, which are favourable for settlement. River terraces, gentle slopes, flat drained-out lake bottoms, a stable cone-like feature made by accumulation of eroded material from above, glacial terraces and glaciated basins form areas where villages are located. Availability of agricultural land, good soil and easy supply of water for drinking and irrigation are other physical factors. Among cultural factors are religious centres, temples, route centres, passes, river junctions and Chattis and other centres along the routes.

Similar terraces occur in the Tons, Giri and Yamuna valleys. There are good terraces below Jubbal on the Pabar river, and along to Tiuni an the Tons and further down. In the Yamuna valley terraces are well developed near Barkot, which is situated on a double terrace. The village of Kuthnaur is located on a hanging terrace. In the Bhagirathi valley, very well developed terraces occur from Bhatwari to Tehri. Villages are located all along these terraces, and the road to Gangotri passes along them. The terraces are covered by fertile cultivated fields.

In the Alakananda valley terraces spread from Pipalkothi to Kirtinagar. The terrace of Gauchar is notable. Also the wide terrace at Srinagar is very well developed. There are terraces in the Sarju valley at Bageshwar, Kapkot and Vani. Throughout Nepal terraces occur in all important valleys. In the eastern Himalaya terraces occur in the Tista and other valleys.

Instances of terraces can be multiplied. They provide a linear settlement pattern developed along rivers like chains connected by roads and tracks. While the flat terrace lands are cultivated, the houses are located on their fringes on higher slopes, or in hollows cut by streams.

Exposed Lake Bottoms

A number of flat basins are found in the Himalayas, which are exposed lake bottoms. With their level topography and fertile soils, they are densely populated. The most notable of such basins is the valley of Kashmir. It is a densely populated patch with numerous villages scattered all over the area. Villages are located on mounds. The valley is surrounded by terraces like formations of lake deposits called Karewas. On their top villages are found at places where irrigation is possible. Also villages are found at the bases of Karewa cliffs, where streams come out of them, such are the villages of Martand and Tanmarg.

The Vale of Kathmandu in Nepal is of a similar nature. It is the most densely populated area in Nepal. The settlement pattern is similar to that in Kashmir. Villages are located near springs found at the junctions of sand and clay layers.

The Vale of Pokhra in western Nepal is also an exposed lake bottom twelve kilometres by six kilometres in size. Here is the big village of Pokhra with its wide tree-lined avenue nearly eight kilometres long-. It is lined by two streams on two sides and a raised platform in the centre for weary travellers to sit and rest-Shops and houses line both sides of the avenue. Temples decorate the centre of the avenue. It is an ideal settlement pattern of a flat-bottomed valley.

The Vale of Manipur is another densely populated lake bottom in the eastern wing of the northern mountain wall. Examples of other densely populated flat-bottomed basins are the bowl of Baijnath below Kosani, the Mandi basin in the Beas valley and the Baspa basin in the Sutlej valley.

Glacial Terraces

These terraces occur at an altitude of three to four thousand metres, and are utilized for locating villages. The big village of Kharsali in the Yamuna valley is a good example of such location. The Kedarnath temple and the village attached to it are also situated on a large glacial terrace. The Thyangbochė monastery is located on a glacial terrace below Mount Everest. Glacial terraces are also used as temporary settlements. The terraces at Zojpal, Sheshnag and Panchtarni on the way to Amarnath are used for such purpose.

Glaciated Valleys

These valleys are found below the glaciated terraces. Their shape is like a U. Here settlements are found near the flat bottoms of valleys. The villages are connected by routes along the valleys, which are enclosed by high cliffs, produced by the truncation of spurs by past glaciers. Above the spurs are found shoulders of gradual, gentler slopes. Here again villages are situated. Cultivation is practised by terracing the slopes. Sonamarg in Kashmir and Bashisht in Kulu are examples.

Amphitheatres

In the high Himalaya, where glaciation has taken place earlier, amphitheatre-like features are found. Settlements are found inside them. The Badrinath temple and the settlement attached to it are found in a giant amphitheatre. The village of Mana slightly above this place is also situated in a bowl-like feature. The village of Beef just below Kharsali in the Yamuna valley is also located in an amphitheatre.

River Confluences

In the higher Himalayas, much debris is deposited at river junctions so that a flat triangular area is produced. This can be irrigated easily. Such places become good spots for settlements. The best example is that of the big village of Padam in the Rupshu valley in Jammu and Kashmir. Here again the attractive resort of Pahalgam is located on a triangular shaped flat area at the junction of the Arau and Sheshnag rivers. Hanuman Chatti on the way to Badrinath is near the triangular flat area at the junction of the Khirao Ganga and the Alakananda. Many other examples can be cited.

River confluences are also supposed-to be sacred, and are called Prayags, where settlements are often found. Such spots are supposed to be holy. Temples are located here and people settle at the junctions of rivers. Devprayag is situated at the confluence of the Bhagirathi and the Alakananda. Rudraprayag is located where the Mandakini joins the Alakananda. Again Karnaprayag is located at the junction of the Pindar and the Alakananda. Vishnuprayag is situated at a point below Joshimath where the Dhaoli Ganga meets the Alakananda.

Water Supply

This is a very important factor for the siting of villages in the Himalaya, where there is often scarcity of water. Sites near springs are very common. In the Vale of Kashmir, villages are found near voclusian springs. They are giant springs coming out of limestone caverns, such as those at Verinag, Anantnag and Martand. Mention has already-been made of spring locations in the Vale of Kathmandu at the junctions of sandy layers and clays. Masonry tanks are built around the springs with crocodile or lion heads discharging water. Springs are located in a line between

Dagshai and Sabathu in the Simla Hills. Villages are located along this line called the pipe line. Water is collected by a pipe and supplies water to Sabathu, Dagshai and Solan.

Hot Springs

They also provide sites for villages in the Himalaya as waters are health-giving. Also hot water is good for bathing, washing clothes and cleaning pots. Bashisht village in the Kulu valley is a good example. Gangnani in the Bhagirathi valley, Gouri Kund in the Mandakini valley, the Badrinath temple itself and the temple of Yamunotri and Janaki Chatti are all near hot springs. The hot springs may be charged with sulphur. They may be also radio-active. Muktinath in Nepal north of Dhaolagiri is a well-known hot spring. Often religious spots are located near spring§, such as Yamunotri and Badrinath, already mentioned. Near Jammu, the temple of Vaishno Devi is situated near a spring.

Sites that are free from Natural Calamities

Villages are located in places which are not prone to suffer natural calamities. Areas troubled by landslides, rock-falls, floods, avalanches, etc., are avoided. The situation of the temple of Badrinath is not good in one way, as it is situated below an avalanche slope. It has been destroyed now and then by avalanches falling from the higher slopes of Narayan Parvat. The settlement is now being developed on the other side of the river over which a new bridge has been constructed. The temple cannot be shifted on this side as it stands upon a hot spring, and the site is considered sacred.

Village Kamru situated on a sharp ridge-top in the Baspa valley is quite safe from avalanches. The situation of Chini is similar. Village Chaura at the border of the Kinnaur district is also located on a sharp spur-top.

Twin Villages

The villages Talla and Malla in the Milam valley have already been mentioned. Even the resort of Nainital has twin quarters, Talli and Malli. Semi-nomadic people often have such villages. The village of Barari belonging to the Jadhs in the Bhagirathi valley may be taken as the twin to Nilang village higher up. Many Jadhs have two houses, one in Barari and the other in Nilang.

Junctions of two Physiographic Regions

Many settlements have grown up in such an environment—Kathgodam, Kalka, Koilabbasa etc., may be taken as examples of such locations, where hills meet plains. The location of Jammu is also of the same type.

Sites of Cultural Importance

Such sites are found along routes, passes, road junctions, river-crossings, etc. Settlements also grow near sacred spots such as Gangotri, Kedarnath, the Hemis monastery and so on.

Urban Settlements

There are few urban settlements in the Himalaya of any importance. They may be divided into a number of types.

Modern Hill Resorts

Such towns are modern and were developed by the British, who liked to escape the heat of the plains during summer. The Governors and Viceroys shifted their offices to the hills in summer. The families of Britishers lived in hill stations. Many cantonments were attached to hill resorts where British soldiers lived to avoid heat. The locations of such towns do not depend much on topographic factors.

Simla, the biggest of such hill stations, is situated on a sharp ridge joining two peaks, Jako and Prospect Hill. The city spreads on the gentle, southern sunny slopes. It is reached by a hill railway and a road, nearly eighty kilometres long. Water supply is difficult, but it has to be arranged by pumping up water by electric power.

Movement in the town is along steep up and down roads. The main advantage is the low-temperature due to altitude. Other hill stations are also located in this area, such as Kasauti, Solon, Dagshai, Sabathu and so forth. Dalhousie was also a big hill station developed by the British. But after the partition of India, its extreme situation has made it less important. Dharamsala under the shadow of snowy Dhaola Dhar is another British hill resort.

Mussoorie, called the queen of all hill stations, is located on a 2,000-metre high limestone ridge rising abruptly above the Doon valley, from where it is reached by a serpentine road. Nainital is a British-made hill station with a difference. It is located in a depression filled with a blue lake. But Ranikhet near it is situated on a ridge-top. Both have cantonments attached to them. The biggest hill resort in the east is Darjeeling developed on a sharp spur, it has a protected position behind the Ghoom ridge. Shillong in the eastern hills is spacious, as it spreads over a rolling plateau. It is the capital of the new state of Meghalaya.

Old Towns and Princely Capitals

Old hill towns were located very often on wide terraces which provided space for all ideal location for the growth of the town. Food could be grown in adjoining terraces. Many small Himalayan kingdoms had their capitals located in such places. The sites are definitely better than the sites of British towns.

Chamba is situated among wide terraces of the river Ravi; Kulu, is located on a wide terrace of the river Beas. Bilaspur was again situated on a terrace of the Sutlej. But it is now drowned in the waters of the Govind Sagar.

Old capitals of kingdoms are also found in exposed lake bottoms. They have developed as cultural centres of the kingdoms. Srinagar is the capital city of Jammu and Kashmir. It is also the centre of Kashmiri culture. Similarly, Kathmandu is the capital city and cultural centre of Nepal. Again, Imphal in an exposed lake bottom is the centre of Manipuri culture. Some towns are located in bowl-like water gathering basins at the sources of streams, which were perhaps glacial amphitheatres in the past. The towns of Pithoragarh and Champawat in Uttarakhand, and Jubbal in Himachal Pradesh have such situations.

The Optimum Belt

The most important factor in the development of rural and urban settlements in the Himalaya is altitude. Even the various favourable topographic land-forms have some relation with altitude. The following altitudinal zones may be easily recognized.

(i) *0 to 1,000 metres :* Here the topography is rough. There are deep gorges and sharp water partings. The area is covered with tropical forests infested by wild animals including elephants and tigers. The climate is warm and humid. There are few settlements of any importance.

(ii) *1,000 to 2,500 metres :* This may be called the optimum belt in the Himalaya. River terraces are very well developed. Upon them are located villages and towns. The climate is cool and comfortable. Cultivation can be easily carried out on terraces and gentle slopes. There are fewer forests. The majority of the people of the Himalaya live in this belt.

(iii) *2,500 to 3,000 metres:* This belt is extremely rugged. It is cooler and is generally covered by coniferous forests. But there are some villages with cultivation in favourable Places. River gorges are deep and often U-shaped cliffs of truncated spurs and thundering waterfalls provide awesome grandeur.

(iv) *3,000 to 4,000 metres:* This belt is inhabited by semi-nomades, who are shepherds and traders. There is much grazing in alpine meadows. There are some temporary summer dwellings. The few settlements are found on glacial terraces and amphitheatres.

(v) *4,000 to 5,000 metres :* This belt is cold, rocky and snowy. There may be a few pastures in favourable spots. There are generally no permanent settlements. Only Seminomads visit there areas now and then.

(vi) *Above 5,000 metres :* These are ragged empty regions of rock, ice and glaciers.

9

Himalayan Paryavaran

Darjeeling and its adjoining areas are known for tea. The presence of tea in this region has given it a special identity. But has also left the regional economy vulnerable due to pronounced sectoral dependence on one capitalised sector of the rural economy Relation of development with tea in Darjeeling has thus been a seesaw. Investigating rural problems of areas where the health of the tea industry has declined, the paper makes out a case for community development as a viable alternative.

The Demography of Darjeeling

Darjeeling district at the northernmost tip of West Bengal comprises 1.96 per cent of total land area of the state but 3.55 per cent of its population, Two blocks in the plains and eight hill blocks constitute the district, and as such the hill area is around 2436.55 sq. km. or 77 per cent of district area. 53 per cent of the district's population of over 13 lakhs lives in the hills. Altitudinal variation over the region ranges from 500ft. to more than 10000 ft., presenting widely differentiated topographical conditions.

To understand the character of economic development and its impact on socioeconomic equilibrium in the hill region, demographic features must first be taken into account. Rapid increases in population impose tremendous pressure on land and other natural resources, disturbing the economic and ecological order in the hills. Demographic pressure leads to encroachment on forest and pasture lands, and puts immense burdens on forest reserves, which have to sustain both timber and fuelwood requirements.

Increasing Wood Demand

With added entry of wood-demands from the plains, commercial exploitation of forest resources takes place at a rapid rate. Per capita forest land in the district was estimated as 12 ha, in 1971. However if the present rate of demographic increase and exploitation continues, this would dwindle to 0.075 ha. by the year 2000 (Sankrityayana, 1996). Demographic pressure also leads to scarcity of economic opportunities, which are already limited by the natural constraints in hill regions.

Growth of population has traditionally been high in the Darjeeling hills. The problem is further compounded by immigration from across the state and across the border. Total population in the hills of the district rose from roughly 1000 in 1850 to 173342 in 1901, registering 17-fold increase over the initial fifty years. Over the next half-century, it nearly doubled to stand at, 28725 in 1951. Between 1951 and 1991, the hill population doubled once more rising to 684818. The impact of the high growth rate of population is not however uniformly dispersed which will be clear from the following table.

Table 9.1 : Block Features in Darjeeling District

Development Blocks	Towns	Villages	Urbanisation Rate % 1981	1991	Population Density/sq.km 1981	1991
Hills						
Darjeeling Pulbazar	1	81	38.6	52.5	860	1102
Rangli Rangliot	-	20	0	0	156	214
Sukhiapokhri Jorebungalow	-	81	0	0	402	478
Kurseong	1	64	35.6	36.2	162	210
Mirik	-	13	0	0	75	200
Kalimpong I	1	28	38.0	42.0	161	167
Kalimpong II	-	70	0	0	125	170
Gorubathan	-	36	0	0	181	176
Plains						
Siliguri Naxalbari	2	222	43.6	47.8	1271	1548
Kharibari Phansidewa	-	485	0	0	349	448
District Total	**4**	**1104**	**27.9**	**30.5**	**325**	**413**

Source: Compiled from the Darjeeling District Statistical Handbol. (1994) and Lead-Bank data for the Central Bank of India.

Population Densities

Comparison of population densities provided in the table show high settlement densities in tea-blocks to the west of the Tista, which include Darjeeling-Pulbazar, Rangli-Rangliot, Sukhiapokhri-Jorebungalow, Kurseong and Mirik, and considerably thinner settlement in the eastern hill-blocks. Population density has also increased substantially over the 1981-1991 decade in the tea-block while remaining steadier in non-tea hill-blocks. Unequal demographic pressure is therefore exerted on tea-blocks, including a substantial component of inter-block migration which has been mounting over the years. Density is highest in the Darjeeling-Pulbazar block since the town of Darjeeling is located within it. The demographic pressures on this block bear the character of rural-urban migration, reasons for which will be analysed later.

Demographic Interlinkages with the Economy

Density differences in the hills arc closely associated with the differential land-use pattern. As earlier stated, inter-block land-use differences locate the tea industry in specified blocks to the west of the Tista, with Mirik accounting for the highest tea-area in ratio-terms, followed sequentially by Darjeeling-Pulbazar, Kurseong, Sukhiapokhri-Jorebungalow and Rangli-Rangliot blocks.

Lands in the twin Kalimpong blocks and Gorubathan share similar environmental and ecological conditions, out are confined largely to agricultural use when not left under forest, except for 6 tea estates located in the last-named block. Gorubathan incidentally also has highest ratio of forest cover at 63.3 per cent.

Forest cover in most other blocks is well below desirable levels, particularly it the tea-tracts where unculturable areas have long been denuded of their forests by managements and workers intent on securing firewood. In face of rising population density and precommitment of lands to tea, hidden but acute land-shortages prevail in the Darjeeling hills-particularly in tea-tracts. Characteristics and consequences of rural economic problems in the Darjeeling tea-tracts are therefore vastly different from those in the agricultural blocks of the hills or elsewhere.

Table 9.2 : Block Land-use Patterns in Darjeeling District

Blocks	*Land Allocation* Tea Estates	% Tea Area	% Forest Area	%Cultivated Lands	% Waste Lands
Hills					
Darjeeling Pulbazar	22	361	7.1	46.0	43.1
Rangli Rangliot	8	13.4	6.5	7.9	69.8
Sukhiapokhri Jorebungalow	28	23.7	59.1	8,0	1.0
Kurseong	26	31.4	48.9	27.0	n.a.
Mirik	12	68.4	18.5	12.7	62.3
Kalimpong I	-	-	4.2	90.4	4.8
Kalimpong II	-	-	24.2	50.4	24.0
Gorubathan	6	6.9	63.3	12.7	23.8
Plains					
Siliguri Naxalbar	25	30.6	28.0	45.2	43.2
Kharibari Phansidewa	18	14.7	1.2	61.5	37.3
District Total	**145**	**20.0**	**34.3**	**41.6**	**n.a.**

Source: Compiled from the Darjeeling District Statistical Handbook (1994) and Lead-Bank data for the Central Bank of India.

The Origins of Tea

The demarcation of land-use in the Darjeeling hills in terms of tea and non-tea areas was made for purely historical reasons by the British at the time of the leasing out of lands to the tea industry under the Waste Land Acts. The western land-tracts of the district had already been incorporated to British India by 1865 when tea plantation activity began. Kalimpong and the other eastern blocks were annexed later to the district from Bhutan and were thus identified as khasinahal, with cultivation as the mainstay of their resident population.

Migration patterns and settlement densities in the two regions were also greatly influenced by the differences that evolved in the character of the predominant economic activity. Selective location of the tea industry exerted a strong economic pull on migrant labour to settle the west of the district. Opening of new estates created employment opportunities which were largely responsible for traditionally-high population densities in the western hills. The blocks cast of the Tista with their agrarian character of subsistence cultivation did not attract as many migrants-a feature that still continues-because of limited ability to absorb additional labour.

Tea Industry

Over the first half of the present century, he tea-industry showed consistent performance with rising outputs and profits, with a large part of the latter being ploughed back for further development of the industry. However with the end of expansion of acreage under tea, new employment opportunities in the sector dwindled and the stagnation and sickness of tea-estates that followed rendered a sizeable number of plantation workers jobless. One of the features of the tea-industry in the post-Independence period was the gradual acquisition of the estates by Indian entrepreneurs, and in several cases, bad management coupled with the desire for quick profits and the lick of adequate commitment cut towards long-term development of the industry by the new owners are held responsible for this state of affairs.

Instances also exist (Chaudhuri, 1986) where surpluses generated in the tea business were invested in other industries, resulting in the gradual development of in economic crisis in the hills. Continuance in such trends will lead to more and more hill tea estates becoming sick in the near future. In general, the stagnation of the tea industry has led to stagnation in the hill region, particularly of rural areas in the tea-blocks where people do not have any alternative means of livelihood. A rural economic crisis therefore looms over the district.

The Rural Crisis

The economic situation in the non-tea areas is not materially different, since difficult topographical conditions ensure perpetuation of the subsistence mode of cultivation. Agricultural production therefore remains at a very low level with depleting soil fertility and the prevalence of monocropping because of absence of the irrigation facility. Traditional agriculture in the hills cannot generate the level of surplus necessary for economic development. Cultivators, moreover, cannot secure remunerative prices for what surplus they do produce because of isolation of the rural economy from urban centres and the resulting inaccessibility of urban markets. This affects the earning power of the rural population.

Survival Needs

This has its effects. The survival needs of impoverished rural communities often lead to human entry into the forests, and illegal felling and timber-smuggling, resulting in rapid decline of forest cover, which aggravates soil erosion and other environmental problems. Besides the rural need for fuel and fodder, wood-demands from urban areas and the plains have also been an important factor in the massive forest loss in the Himalaya. A, study by the Tata Energy Research Institute [TERI] (quoted in PCED, 199 7) shows that fuelwood consumption per day rapidly rises from 1.03cum. in the plains to 2.12 cum. at 600m. altitude and to 2.49cum. at 2000m. in the face of poverty and the lack of access to commercial energy sources like coal, kerosene, LPG. and so on.

Another section of the rural population, finding no employment in the villages, moves into the hill towns to engage in any of the occupations associated with tourism. However, with Darjeeling being the central focus of tourism, the industry remains highly centralised and its effects area-specific. Thus, instead of dispersing economic opportunity over all regions land all sections of people in the hills, the benefits of the industry remain largely confined to the hill township of Darjeeling and to the small number of people involved in the hotel business. Most people engaged in tourism-related activity are seasonally employed and remain practically jobless through the remainder of the year. Dissatisfaction among the partially employed or unemployed rises.

The Urban Crisis

When Darjeeling was acquired by the British as a potential summer resort, it was practically uninhabited. The process of developing it into a 'hill-station' commenced around 1835. Over the next 15 years, the town had grown sufficiently for the authorities to consider it reasonable to set up a municipality to look after the provision of civic amenities. Darjeeling Municipality was thus set up in 1850 when population of the town was around 10 thousand, and civic and infrastructural facilities were therefore designed to cater to the needs of a limited population. Population in the township now exceeds 80 thousand and its pressures already exceed urban carrying capacity.

Floating Population

Added to this is the large floating population of tourists. It brings with tremendous load to bear on water resources, atmosphere, vegetation and human settlements. For water-supply, the town depends almost entirely on the Senchal lakes which lack capacity to provide even the volumetric water requirements of the local population. Thus the influx of seasonal tourists raises the problem to crisis dimensions. The increased population associated with tourism also produces a growing volume of garbage and sewage, the disposal of which is fast becoming an area of major concern. Tourism also brings increased transportation activity and traffic, creating the new problem of environmental pollution.

Increased human activity on the creation of more tourist facilities has a crucial impact on the fragile ecology of the hills because of the destruction of natural vegetation, which has now proceeded to such an extent that even rainfall of 25 mm can cause landslides. Unplanned expansion of tourist accommodations, resorts and shopping areas constantly displace the local population and consequently lead to a severe curtailment of housing amenities. Urban development of the pattern followed by Darjeeling in recent years can hardly be sustainable from the standpoint of the local people.

The Crisis of Development

Economic development in the hill regions of Darjeeling district has therefore been unplanned, uneven and skewed, producing sharp cleavages between richers and poorer and rural and urban sections of the community. Moreover, development of this kind lacks an eco-friendly character and thus cannot be sustained into the long run.

While a lot of attention has been devoted in recent times to urban development in Darjeeling with the i of tourism and related ecological questions coming' for scrutiny, unfortunately all such efforts fail to address the real issues of development in the villages where, per cent of the hill population resides. In the urban-one debate, the stagnation of the village economy and; the despair of the impoverished rural community are, properly represented. Economic and social equilibrium of the hill region cannot be restored without creating sufficiency in the rural economy. Development intervention—whether sourced from government or non-governmental agencies-into the economics of the system is therefore necessary to set the village economy into motion again.

However, since both mountain settlements and mountain communities are widely differentiated in terms as geographical and sociocultural entities, no central model of rural development would be meaningful. Development of local communities through their participation in the decision-making process is probably the most viable alternative to resolve the present rural crisis. It is also capable of letting the fruits of development trickle down to the poor.

Case Study

Community development along such lines has, initiated in three tea-villages by the Regional Community Development Committee, [RCDC], a non-governmental organisation instituted by the Darjeeling Jesuits of Bengal. The villages covered, namely Harsing, Dabaipam and Yankhoo Busties, lie on the Lebong Spur of the Darjeeling-Jalapahar Range. Harsing Busy is situated 10km. from Darjeeling town, and Yankhoo and Dabaipani further away across the valley. The busties (villages come under the Darjeeling (Sadar) Sub-division, and within the jurisdiction of the Block Development Bijanbari.

Asked about the objective of the programme, launched in these villages, Mr. C.B. Rai, Convenor, RCDC (also Principal, St. Joseph's College, Darjeeling) remarked: "We are trying to organise the village communities so as to make them self-confident, self-reliant and capable of developing their potentialities." The most powerful instrument of community development in these three villages is the cooperative movement. To understand the impact of the programme, it is first necessary to examine the socioeconomic background of community life in the three villages.

Table 9.3 : Demographic Features of the Tea Villages

Village	*Families*	*Population*			*Literates*		
		Males	*Females*	*Total*	*Males*	*Females*	*Total*
Harsing	103	250	216	466	57.4	37.5	48.2
Dabaipani	149	426	334	760	89.3	71.0	81.3
Yankhoo	55	130	113	243	84.0	56.1	72.3
Total	**307**	**806**	**708**	**1469**	**77.4**	**56.9**	**68.2**

Source: Compiled from RCDC Field Investigation Report (1996).

The members of the three villages belong mainly to the Rai, Chhetri and Magar (Mukhia) sub-communities among the Nepalis, all speaking Nepali as their mother-tongue. Average family-sizes computed on the table for Harsing, Dabaipani and Yankhoo villages are 4.5, 5.1 and 4.3 respectively. There is substantial shortfall in female-male ratios [FMRs] for all villages at 864 (Harsing), 784 (Dabaipani) and 869 (Yankhoo), compared to FMR for Darjeeling district at 934 (rural) and FMR for India at 929. Overall literacy per centages for the community of the three villages as a whole consistently exceed district averages of 67.07 per cent (male), 47.07 per cent (female) and 57.95 per cent (combined).

Literacy in the Region

However literacy is lower in Harsing than in the other two villages and is commendably high among both males and females in Dabaipani village. Rural literacy patterns in the rural community are more or less in keeping with the higher literacy attainment of the Darjeeling hill which exceeds state and national averages and breaks the archetype of the poor illiterate villager, substituting it instead with the archetype of the poor literate villager typical to the rural community of the Darjeeling hills. Literacy attainments however do not reflect educational attainment since school-dropout rates in the villages are very high, with little spread of higher education due to poor economic status of the village dwellers.

With the villagers being poor otherwise, their relative economic position is mainly determined by their holdings of land and livestock assets which are the subject of the next table. Since average size of the holdings is extremely small in all three villages, prospective incomes from cultivation are also very low. Livestock-holding per village family is small due to the inadvisability of rearing animals in the face of unremunerative returns on them, to which geographical isolation of the villages, lack of proper marketing facilities and high mark-up profits extracted by middlemen all contribute. The low holding of livestock by the village residents also reduces their nutritional status in the long run.

Table 9.4 : Economic Assets of the Tea Villages

Assets	*Harsing*	*Dabaipani*	*Yankhoo*	*Total*
*Land (acres)**				
Total Holdings	130.25	22.22	51.45	401.92
Average Holding	1.26	1.80	1.25	1.51
Livestock (heads)				
Milch	88	78	21	187
Non-milch	30	94	33	157
Calves	28	47	14	89
Total	148	219	68	433
Cattle-heads				
Pigs	4	22	9	
Goats	179	68	25	272
Chickens	153	101	55	309

* excl. unascertained landholdings of 22 households in Dabaipani & 14 households in Yankhoo villages.

Source: Compiled from RCDC Field Investigation Report (1996).

The Source of Crisis

Tracing the history of the three villages, it is that observed that Harsing, Dabaipani and Yankhoo Busties once formed the bulk of Harrison Tea Estate. Residents of the villages mainly comprise erstwhile tea-labourers who once drew their livelihood solely from the vanished estate and were not therefore engaged in other productive activity. Agriculture as a means of livelihood was thus alien to them because of their long association with wage-employment in tea as plantation-labour. The economic equilibrium of the three villages was disturbed when the tea estate closed down in 1952 because of mismanagement and irresponsible trade unionism, with the result that an acute crisis surfaced in village life. Survival needs first dictated to the village residents that they pluck tea leaves from bushes in the closed garden over the plucking season and sell them to neighbouring estates.

In addition, they also started felling trees in the tea-garden reserve forest, seeking to survive by selling firewood and charcoal. This resulted in gradual depletion of forest cover leading to soil instability, water crisis and environmental disturbance in the area. These unregulated continued for 10 years. Finally in 1962, when the reserve forest was exhausted and the possibility of the reopening of the estate grew progressively bleak, people began to uproot tea bushes to cultivate other crops like maize and millet.

Crop Productivity

However it turned out that the total production and productivity of these crops were very low because of the lack of experience and cultivation-knowledge on the part of the people and due to the soil having been rendered highly acidic by cultivation of tea on it over a long duration. Because of their previous tea-labour identity, it was also much harder for the village residents psychologically to accept the status of agricultural cultivators. People living in the erstwhile tea-villages therefore largely failed to themselves into agricultural communities.

Again their status as ex-tea labour militated against their taking tip petty trading enterprise because of the lack of entrepreneurial spirit and working capital. People of the three erstwhile tea-villages therefore continued to linger at subsistence levels. No assistance came to the villagers from Government because of the tea-garden status of the area. Labour-welfare in the tea estates is regulated by the Plantation Labour Act, 1951. That no Government welfare schemes have ever functioned in these three villages even after four decades of the garden's closure is in itself a surprising event.

Struggles for Subsistence

In absence of any assistance from government, village residents sought to supplement their in meagre income from sales of milk. To finance the purchase of cows for this purpose, they secured loans at highly usurious rates of interest (varying between 72-120 per cent per annum) from village middlemen, with the added liability also of having to sell milk to their creditor-middlemen at lower-than-market prices.

Exploitation by Middlemen

A major share in the profits from this new enterprise was therefore extracted by these middlemen both by way of high rates of interest and undervalued milk-sales. The lack of remunerative prices for milk later discouraged a number of village families from keeping cows, which has lowered availability of organic manure and has further impoverished the villagers through the consequently-lowered agricultural productivity. In 1985, average sales of milk per day were 600 litres for 135 families. By 1996, this had declined to daily sales of 502 litres for 307 families, indicating the declining number of cows in the three villages. It had been noted earlier that the people of Harsing, Dabaipani and Yankhoo Busties could not adapt to agriculture as their means of livelihood because of their long prior association as wage labour with the tea estate.

The same mentality lingers today, and whenever work is available, they shift to wage labour at construction sites or in agriculture, even when they know that such work is temporary. Nevertheless they prefer performing wage labour to being agriculturists. Retention of this mentality is bound to tell on the nature of agricultural production and productivity in the villages. Since landholdings of the villagers are small and uneconomic (Table 9.4), they yield very low income.

Thus the average annual income per village household is estimated at Rs. 11822.76 or less than a thousand rupees per month per family. Cultivation activity in the three villages mainly comprises subsistence cropping where maize, millet, vegetables and other cereals are produced by the villagers mainly for their own consumption and not for outside sale. The main cash crops are ginger, cardamom and oranges; however the productivity of these in villages under study is very low, since production of cash crops is highly input-intensive and yields are drastically reduced when inputs cannot be applied in sufficient quantity. Poverty of the village people renders it impossible for them to bear expenses on inputs at a level that would ensure good returns.

Almost all households in the villages also collect amlesho which has high commercial value. Amlesho is a multipurpose plant whose leaves are used as fodder, whose woody stems are used as fuel and whose blossoms are used to make brooms. Most families at Dabaipani and a few at Yankhoo have tea bushes surrounding their fields. They therefore also pluck tea leaves from February to November, hand-process them into tea which they then sell to the market to supplement income.

Income Sources

Looking now at sale-values of commodities produced by the village residents, milk sales are seen to occupy prime position, even though sale prices are hardly remunerative at between Rs. 3-5 per litre in the village, at a time when retail market prices range between Rs. 8-10 per litre. The commodity placing next in order of if commercial importance to the village residents is ginger. Again the prices generally realised by the villagers at around Rs. 8.75 per kg. are well below the prices eventually realised in the market by ginger traders at around Rs. 20 per kg. Prices for hand-processed tea in the villages are reported to be around Rs. 65 per kg. for the same tea leaves which fetch the trader between Rs. 130-150 per kg. in the Darjeeling market. Village residents therefore fail to secure good prices for each of the commodities they sell and the process of income-augmentation is considerably slowed down.

Table 9.5 : Income per annum from Commodity Sales in the Tea Villages

Commodity	*Harsing*	*Dabaipani*	*Yankhoo*	*Total*
Milk				
Quantity (litres)	87235	72270	23725	18330
Cash Realisation	361715	237980	79890	679585
Cardamom				
Quantity (kg.)	105	187	210	1502
Cash Realisation	3885	43919	7770	55574
Ginger				
Quantity (kg.)	6420	60220	5553	72193
Cash Realisation	56175	526925	48588	631688
Amlesho				
Quantity (kg.)	920	5733	1000	7653
Cash Realisation	9200	57330	10000	76530,
Orange				
Cash Realisation	25000	12000	-	37000
Tea				
Quantity (kg.)	168	3418	185	3771
Cash Realisation	10920	222170	12025	2451115
Total Cash Realisation from all Commodities	466895	1100324	158273	1725492

Source: Compiled from RCDC Field investigation Report (1996).

Hidden Exploitation

Adequate market facilities are a vital prerequisite to overall development of agriculture. Unfortunately, market facilities in the Darjeeling hills are woefully inadequate because of lacunae in the communication networks that connect the villages. Inaccessibility of markets reduces the earning capacity of the rural people, who are heavily dependent on middlemen for the marketing of produce. Most of these middlemen are village shopkeepers who advance production-related credit to village producers on stipulation that such borrowers will repay their loans by selling agricultural produce to the lenders at prices dictated by them.

Cost of Loan

An element of implicit or invisible interest is added to the visible interest cost of the loan by the undervaluation of products. Rural credit transactions are therefore linked with the marketing of agricultural produce and arc responsible for the extraction of an extra margin of surplus from the rural economy. Indication of the extent to which such 'surpluses' are realised by middlemen from the village producers of Harsing, Dabaipani and Yankhoo is provided in the following table for three commodities of market sale.

Table 9.6 : Extraction of Hidden Surpluses by Traders through the Marketing Linkage

Commodity	*Annual Sale*	*Value Received*	*Market Valuation*	*Surplus Extraction*
Milk	183230 lts.	679585	1649070	969485
Ginger	72193 kg.	631688	1443860	812171
Tea	3771 kg.	245115	565650	320535
Hidden	Milk	Ginger	Tea	
Surplus per unit	5,29	11.25	8550	
Commodity	per lt.	per kg.	per kg.	

Source: Compiled & computed on RCDC Field Investigation Report (1996).

For the middleman-trader therefore, direct trade with the village producer is a highly profitable enterprise, while the extraction of the huge amount of surplus involved from the rural economy results in further impoverishment of rural community. Such extraction of hidden surplus through market linkage is not unique to the hill economy. Another study in Cooch Behar in 1990-91 by the present author confirmed the presence of identical trader-based linkages. However, in the hill regions, the scope for agricultural progress is necessarily more limited than in the plains as market inaccessibility is greater. Thus the impoverished rural community fails to improve its lot and is compelled to remain at subsistence levels.

Income Profiles

Income is therefore generally low in all three villages, and even between villages, there are disparities in income levels. In terms of income per household, Yankhoo Busty is worst-placed among the three lea villages, while Harsing Busty is best-placed. Various factors might be causally attributed to explain the lowest positioning of Yankhoo in terms of income. The village experiences the greatest isolation from the market and therefore suffers most in terms of unremunerative prices. Land in the village is rocky and therefore less suitable for cultivation which also explains the lower levels of settlement in the village. Cash cropping is low due to unfavourable topographical conditions which prevent the village from growing oranges. Reduced number of cattleheads in recent years is another factor leading to low income in the village. The highest income at Harsing is mainly attributable to its relatively closer proximity to Darjeeling and the market.

Table 9.7 : Household and Per capita Incomes in the Tea Villages

Village	*Income per Annum*	*Income per Family*	*Income per Capita*
Harsing	834367	14899.41	2799.87
Dabaipani	1352288	1136316	2055.15
Yankhoo	155252	6750.08	1008.13

Figures in Rupees.
Source: Compiled from RCDC Field Investigation Report (1996).

The First Intervention

The first attempt to revive the economy of these villages was made in the Severities by the Hayden Hall Institute [HHI], an organisation of the Darjeeling Jesuits addressing social service and community development. With the active support from HHI, the Harsing Busty Dairy Union was set tip in April 1973 with a membership of 25 and a collection of 12 litres of milk. It ran very successfully between 1973 and 1986, with peak performance being recorded in 1979 when membership rose to 135, and 600 litres of milk were being handled each day. In 1987 however, the dairy union collapsed. Although village residents blamed themselves for this event, intrinsic reasons behind the collapse of the union lie elsewhere. "The Union was not developed by people's initiative. It was organised by Hayden Hall and then thrust upon the community", reflects an RCDC activist.

In addition to subsidising the union, MI sponsored various welfare projects such as mother & child care, food for work schemes and paramedical outreach programmes, allowing the impression to develop that all these programmes were part of the dairy union. With suspension of these programmes, the union collapsed. HHI's methodology of intervention "was a top-down approach and people's participation in planning, monitoring and evaluation was not the order of the day. Essentially, the programme had its main focus on economics and did not encompass the totality of development. The ideal of the cooperative as a way of life was never given thought to nor practised" is the final comment of a report.

Reformulation of Community Effort

Holistic planning exercises should evolve from out of the grassroots necessities of people, involving social, economic, political and environmental issues all having direct impact on quality of life. Community-basing these exercises ensures people's participation at every planning stage, and incorporates as an alternative solution, the idea of development through unity at community level,

The analysis earlier has pointed out three major bottlenecks in the rural economic systems of the three tea villages, namely

(i) high rates of interest on loan support and exploitation by middlemen

(ii) lack of institutional credit

(iii) lack of marketing facilities

All three problems needed to be tackled commonly through multipurpose community activity designed, initiated and administered by the people, with minimal intervention from outside. In this concept of "people's" development applied to economic rejuvenation of Harsing, Dabaipani and Yankhoo Busties, highest weightage given by residents was to revival of the milk cooperative. This provided the point of entry for RCDC into the community.

Sanjukta Vikas Cooperative

The Sanjukta Vikas [Joint Development] Cooperative [SVC] was formed as a community organisation with direct participation of the people of the three villages, and commenced operations with 29 litres of milk from 16 March 1997. Initially a milk cooperative, there is a proposal to broaden its base over time to include marketing and other activities. SVC enlisted all village families as its members. Increase in daily milk sales to over 160 litres in subsequent months are a pointer to the success of the cooperative, which moreover offers remunerative pricing of Rs Mitre to milk producers compared to Rs. 3-4/litre they earlier received from middlemen. 14 per cent of earnings from milk sales are deposited in bank accounts of the producers to encourage the saving habit. Over the six-month period between March and September, total milk sales by, SVC have been worth Rs. 2,35,297.

SVC has therefore helped the community by increasing family incomes and freeing the members from the Clutches of middlemen. The success of this experiment in community-cooperation can well be emulated in neighbouring villages as a model for development.

Diversification

SVC also encompasses within its purview a people friendly credit union sponsored for Self-Help Groups by the Darjeeling branch of the Indian Bank. Credit is available up to four times the individual savings deposited by SVC members, who collectively decide the purpose and duration of loans and select recipients. Inbuilt pressure to ensure lean recovery is provided by the condition that no new loan will be advanced to a second recipient until the first loan has been repaid in full. Besides the milk cooperative, diversification of economic activity in the villages is also being contemplated as a means of raising income levels.

Towards this purpose, training on modern farming and dairying practice has arranged for SVC members by the Bidhan Chandra Krishi Viswavidyalaya unit at Kalimpong. Rural-oriented economic activities including the rearing of goals and pigs, floriculture, beekeeping, mushroom cultivation, etc., are also being considered for future inclusion in the community development effort, Widening the scope of community-based development also envisages community healthcare, nutrition and sanitation, and improvement of internal communication in the villages through group labour and material contributions.

Environment Programmes

Water scarcity has been the bane of the tea-villages of Harsing, Dabaipani and Yankhoo, acting as a serious obstacle to raising agricultural productivity. Even drinking water has to be collected after walking for miles, However the scarcity would to a great extent be mitigated if people worked together towards a viable community system of water management, involving the identification, protection and regeneration of water sources, conservation of watersheds and introduction of equitable water-sharing. Much work will have to be done in this direction.

SVC has initiated an afforestation programme, setting tip nursery-capacity to produce 10 thousand saplings. Under the Block Forestry Programme, SVC members have planted 40 thousand saplings of various species in 1997 over 5ha. of degraded land. While species and site selection is decided by SVC, RCDC works in cooperation with it to increase environmental awareness among village residents. 200 smokeless Chulhas capable of reducing firewood consumption by 40 per cent have been introduced so far in the three villages with technical assistance from KVIC, and success in their adoption wills, save fuel and effort besides protecting forest cover in the surrounding hills.

They are made of locally available materials like mud and stone, hence, the design of the Chulhas has been highly cost-effective, and the attention of the neighbouring villages has also been attracted. The village community was also interested in biogas, but the higher initial costs of construction held back its institution. RCDC has however been in touch with the Darjeeling. Gorkha Hill Council [DGHC] over this, which has led to the construction of a biogas pant at Harsing Busty.

The Future of Community Development

RCDC's work with the people has been to raise awareness-levels about developmental and environmental issues confronted by tile community. By community-based planning in which the people are participants, they arc made self-reliant since such planning addresses local needs and stresses the use of resources available locally. Although community work of the RCDC has demonstrated Success in stabilising the rural economy of villages in erstwhile tea areas, the participatory model, followed can be easily extended to working tea-areas, an to non-tea areas where agriculture provides the mainstay, In view of the rural crisis that confronts the Darjeeling hills at all levels, the launching of a 'multipurpose grassroots cooperative' movement may turn out to be tit, best way to resolve the crisis.

References

Buddhadeb Chaudhuri (1986): 'Economy, Migration and Development in the Eastern Himalayas: A Study with Special Reference to the Hill Areas of Darjeeling', in L. P. Vidyarthi, Makhan Jha (eds.): *Ecology, Economy and Religion of the Himalayas*, Orient, New Delhi.

Kanchan Chopra, G.K. Kadekodi, M.N.Murty: *Participatory Development: People & Common Property Resources*, Sage New Delhi.

RCDC (1996): *Field Investigation Report on Harsing, Dabaipani & Yankhoo Busties in Darjeeling*, mimeo.

J. Sankrityayana (1996): 'Region, Location and Sustainable Development: An Examination of the Darjeeling Himalaya', *Himalayan Paryavaran*.

Government:

District Census Handbook, Darjeeling, 1981, 1997.

District Statistical Handbook, Darjeeling, 1994.

10

Development Project Vis-a-Vis Ethno-Religious Sentiments: The Rathong Chu Imbroglio in Sikkim

Introduction

Every welfare state takes up development projects. They are, as the term implies, are meant for the upliftment of society, or loosely one can say, for the material benefit of the common man. The spectrum of such projects include very basic to highly sophisticated infra-structural development. Unfortunately, such projects often create undue confusion among the *hoi polloi*, who become skeptical about the motive of the project planners. Issues produced by confused minds very often prove to be devoid of merit in the long run, but before fizzling out, they may cause imbroglios involving huge wastage of spirit, time and money. The issues raised by the critics of development projects may be as congnizable as conservation of environment and ecology, or as intangible as religious sentiment, tradition and culture.

Very recently, the Rathong Chu hydro-electric project in Sikkim generated much amount of heat and controversy among a section of Sikkimese people on the question of controversy may be cited as a blazing example of development projects in the crucibles of ethno-religious rage. A few years back, some of the leading dailies of the country briefly mentioned about some unrest among a section of people in Sikkim on the issue of a power project. The news item was overshadowed by similar, but, far graver concerns over the Tehri and Narmada Projects in other parts of the country, which involved large-scale submergence of land, rehabilitation of the displaced multitude and above all, the ecological issues. In comparison to the above two projects, the Rathong Chu hydro-electric project was miniscule; neither did it involve submergence of land, nor mass rehabilitation.[1]

Unique Reason

What was unique in the case of Rathong Chu project, however, was the fact that, unlike Tehri or Narmada projects, the unrest in Sikkim involved religion, ethnicity and culture. Here, the agitators were mostly lamas, i.e., the Buddhist monks, representing the 'Sangha' or the Monks' body of Sikkim. All these lamas and their followers primarily belong to the ethnic Bhutia-Lepcha population of Sikkim. This particular ethnic group vehemently opposed the hydro-electric project on the Rathong Chu and brought allegations that the project had been undertaken in total disregard of the religious importance of the area.

It was also alleged that the project was detrimental to the people of Buddhist faith and violated constitutional provisions. The tribal organizations of Sikkim, too, felt that the project may usher in alien culture to their homeland, thus leading their traditional culture towards extinction. So, in association with the monk's body, different tribal organizations of Sikkim raised their voice for the protection of their culture. The ethnic Bhutia-Lepcha population was so alarmed and motivated that writ petitions against the project were filed in the court of law. The peaceful state of

Sikkim saw a unique type of unrest while development took a back seat.

Why Rathong Chu hydel Project?

By virtue of Sikkim's location in one of the most humid tracts of the Himalaya, it has a veritable reserve of water resource. The hydro-electric power potential of this tiny Himalayan state is commendable. It is estimated that the total power potential of Sikkim's water resource is 800 MW, out of which barely 20 MW is presently being tapped.[2] Economists are of the opinion that Sikkim has immense opportunity to mobilise its water resources for domestic as well as commercial purposes. Otherwise handicapped by a highly unjulating terrain, hydel power generation is the only favourable aspect of the mountain topography of the state. Power being regarded as one of the most vital inputs of economic development, there is no alternative but to exploit the water resources which are abundant, renewable and pollution-free.

However, keeping in mind the cost-effectiveness, actual volume of demand, feasibility and the vulnerable eco-system of the region, a judicious planning is called for. In a bid to mobilize the indigenous resources of Sikkim, the Rathong Chu Hydro-Electric Project with an installed capacity of 20MW was conceived as early as 1997, i.e., two years after Sikkim's merger with the Union of India. It may be mentioned here that, at that time the State was producing only 3 MW[3] of electricity indigenously form three mini hydel projects, providing the barest minimum to the handful elites of the State.

Rural electrification was almost unheard of while the State had to buy power from the nieghbouring States to meed the institutional demands. Under such circumstances, the need of the hour was to construct a full-fledged State-owned power project which would not only meet the home demand but also earn some revenue to boost up the stagnant economy of the State.

The Project

The site of the Rathong Chu Hydro Electric Project (RHEP) is located at Yuksam on the left bank of the river Rathong. The gorge through which the river flows is very deep and several hundred metres below the built-up area of the uksam block. The total area involved in the construction of the project is 125 acres, of which 62.5 acres are cultivated land, 25 acres are cardamom plantations and 37.5 acres are forested land.

An area of approximately 4 hectares in the outskirts of Yuksam would be used for project colony, offices, workshops, stores and storeyards.[4] According to the Detailed Project Report (WAPCOS),[5] the RHEP would have simple features comprising trench type diversion weir, power channel, forebay, penstocks and surface power station. The project has an estimated potential of contributing 30 MW of power per year. It would have three generating units, each of 10 MW capacity. The cost of the project, estimated on 1989 price level amounts to Rs. 53.10 crores, corresponding to a cost of Rs.17,690 per KW of installation. It was also worked out that at 9 per cent discount rate, the cost of generation would be 37.10 paise per kWH.

Power Demand in the State

Transmission of power was planned at 132 kV to Gangtok, the State capital, and Geyzing, the headquarters of the West District of Sikkim. The project would be the first independent power plant of the State Government of Sikkim. According to XIII Power survey of the state, the peak demand of Sikkim is expected to grow to 84 MW by 1999-2000.[6] Hence, it is imminent for the State to take up a medium scheme to meet the immediate requirement of

40 + MW. The RHEP is well suited to meet the immediate demand. There are several advantages of the RHEP. Firstly, it would be a run-of-the-river project which requires no reservoir. In absence of reservoir, the problems of submergence and subsequent evacuation-rehabilitation of people do not arise.

Secondly, the requirement of land is minimum and there would be hardly any displacement of inhabitants or deforestation. Thirdly, the capacity of 30 MW matches with the volume of demand. On the whole, the RHEP is viable as well as attractive from the point of view of financial and technological aspects.[7]

The project was planned with a view to selling the surplus electricity to the eastern grid[8] of the country, so that the under-nourished economy of the state would be benefited. The Detailed Project Report of the RHEP was prepared by the Water and Power Consultancy Services in 1989. The report was further examined by the Central Electricity Authority (CEA) and the Central Water Commission (CWC). The project was given techno-economic clearance by the govt. of India in February 1990 at an estimated expenditure of Rs. 71.19 crores.

Thereafter, the clearance of the Ministry of Forest, Govt. of India was obtained in November, 1991 and that of the Ministry of Environment in December 1992. Finally, the State Government accorded administrative approval for its implementation in January 1993.[9] It would be worthwhile to mention here that the construction work of the project does not, by any means, encroach upon the spots significant from the point of view of religion, archaeology or otherwise, neither is it disturbing the bio-diversity of the nearby Kancheandzonga National Park. The private owners of 87.5 acres of land are well-compensated while the forested land of the government is proposed to be adjusted by way of exchange of land as well as replantation after the completion of the construction work. Since the project was first of its kind in the State, due weightage was given to the economic, ecological and cultural aspects.

The Imbroglio

The Rathong Chu Hydro Electric Project, which was conceived in 1977,[10] took about fifteen years to cross all hindrances before the final take off. During the long gestation period, the state witnessed several changes in the Governance, but the proposal survived. The construction work of the project could finally start in the year 1993. But no one was aware of the impending set-back arising from an altogether different quarters. As soon as the construction work started, a section of the Sikkimese people, especially the ethnic Bhutia-Lepcha populatin, demand immediate closure of the project.

Several tribal associations and the Monk's association of Sikkim appealed to the State administration to take steps in this regard. The selection of the project site offended the people of Buddhist faith, who saw it as an encroachment upon their religious rights. The protestations took shape of agitation, followed by hunger strike, heated demonstration and filling of writ petitions in the High Court of Sikkim. In August 1994, three tribal associations namely, the Denzong Lho-Man Choda or the Bhutia-Lepcha Association, the Tribal Women's Association and the Sikkim Tribal Scheme Welfare Association, submitted a writ petition to the High Court.[11] The petitions was written on the matter of "Preservation and safe-guard of Ancient Monuments, Holy and sacred places having Historical Significance".

According to the petitioners, "The said power project would grossly infringe upon the religions sentiments and belief of the State of Sikkim professing Buddhist faith in-as-much-as the said power project would destroy and defile the places of worship, pilgrimage and threaten the existence of very ancient monuments having historical and religious significance to the people of the state of Sikkim which are situated within the periphery of the area where the said project is to come up... The action... amounts to gross violation and interference with the religious rights of the Buddhists of the state of Sikkim. It is evident that the whole argument is based on religious sentiments of the Lepcha-bhutias. The monks of Sikkim proclaim that the area on which the project is to be erected, is the abode of the protective deities of Sikkim. Hence, the area should be protected and kept out of touch of the outsiders even at the cost of development.

The Sentiments

Religion is intricately woven with their day-to-day life that an average sikkimese Buddhist can not think of living unguarded by his deity.' The religious sentiments of the Sikkimese Buddhists are deep-rooted. His belief in spirits makes him extremely vulnerable in the hands of evil power, and he is unable to draw a line of demarcation between myth and truth. There are numerous legends about the mountains, rivers and lakes of sikkim and almost every landmark of nature is associated with certain religious rituals. The Rathong Chu and the yuksam valley is no exception. The annual 'Bum Chu' ritual of tashiding in sikkim is one such event which involves the Rathong Chu.[12]

The Rathong Chu is one of the sacred river to the Buddhist of Sikkim.[13] The river is enlisted in the 'nay sol' or the holy scripture which serves as a guide book of the sacred places of sikkim. According to legends, the river contains some holy sport where the great Buddhist teacher, Guru Padma Sambhava consealed sacred treasures. The river is believed to be blessed by the guru hence, the water of this river is used to refill the sacred vase of tashiding monastery during the annual 'Bum Chu' ritual.

According to the holy scripts, it is the bounden duty of each and every Buddhist to protect the sanctity of the Rathong Chu. Moreover, there are numerous holy places along the bank of the Rathong Chu which must be protected as per the instruction given in the 'nay sol'. The river is revered not only by the Buddhist, but also by the nature worshippers of Sikkim. The valley of vuksam through which the Rathong Chu flows is revered by all Buddhist sikkimese. The valley is like a bowl or a plateau guarded by mountains on all sides. This was the seat of the first chogyal of Sikkim. The 'Norbu Gang' i.e. the coronation throne on which the first chogyal Phuntsog Namgyal was consecrated, is treated as a sacred spot by the sikkimies. The Buddhist of sikkim who are devout followers of the guru's teachings, and tradition-bound to the core, revere the Rathong Chu as well as the 'Norbu-Gang' as the abodes of devine spirits.

The 'Bum Chu' ritual is one of the holiest rituals for the Buddhist of Sikkim. Every year, the Bum Chu, i.e. "water of the sacred vase" ritual takes place at tashiding in west Sikkim on the 14th day of the first month of the Tibetan calendar. The sacred vase is believed to be made of sacred soil, sacred water and five precious jewels collected from holy places by the legendary guru padma sambhava. The vase contains the holy water of the rohtong Chu and is kept under strict vigil in the sanctum sanctorum of the tashiding monastery. Every year, on a full moon day, the vase is opened and three cups of water is taken out of it. The first cup of water goes to the royal family of sikkim, the second cup to the lamas and the third cup of water is distributed among the common devotees who gather at tashiding in large number to receive the holy water on that auspicious day.

Belief

It is believed by the sikkimies Buddhist that a drop of the Bum Chu water relieves the taker from all the distress, evil spirits and miseries of life. They also believe that if the level of water contained in the vase decreases, that is an indication of a bad year for their country. An increase in the level of water, however, symbolizes prosperity and glory. If the water turns turbid, the country may witness untoward incidents, such as war and deaths. Every year, after the ceremonial distribution of holy water, the sacred vase is refilled with water of holy Rathong Chu.

The Sikkimese people, particularly the Bhutias believe that, if the Rathong Chu is harnessed for power generation, the river will lose its sanctity; the water will be polluted and can no longer be used for the 'Bum Chu' ritual. The seed of the imbroglio over the Rathong hydro electricity project lie in the sentiments of a group of people for whom religiousness is second nature. For them, it seems possible to forsake their age-old tradition and belief, even though such practices border irrationality. Even the educated minds are not free from such sentiments. It is rather surprising to note that the persons who took leadership for the cause are educated in the liberated west and belong to the elites of Sikkim.

Ethno-religious Backdrop

Before delving deep into the imbroglio, one should have a look at the ethno-religious background of Sikkim. The population figures of Sikkim, if seen in the light of racial and religious, may produce a deceptive picture. The ethnic mosaic of Sikkim shows the numerical dominance of Hindu Nepalese but does not necessarily reflects the dominance of Hindu culture. Sikkim inherited a predominantly Buddhist lamaistic culture from the theocratic Bhutia aristocrats who ruled the State for more than three centuries.

Despite the diminutive per centage of the Lepcha-Bhutia population, Sikkim's cultural as well as religious Scenes are overwhelimingly dominated by Bhutia tradition and Mahayana Buddhism. Buddhism in sikkim has developed some peculiarities of its own. It has deviated to a great extent from the other forms of Buddhism prevalent in India and elsewhere but has retained the basic traits of Mahayana Buddhism. Here, Buddhism is highly intermingled with animism. The fusion of animism and Buddhism took place slowly but steadily over the centuries and at the moment, it seems and arduous task to demarcate the two.

Buddhists Worship

The Sikkimese Buddhists worship the deities as well as the local spirits who are believed to be inherent in the mountains, rivers, lakes and even rocks, caves and trees of Sikkim. For example, 'Pang Lhabsoi', or the worship of Mount Kanchendzonga is one of the most important religious rituals of the Sikkimese Buddhists. Similarly, they also perform some religious rites with the water of the Rathong Chu, Which, they regard as a holy river. The influences of the 'Sangha', the monks body is very great on the Sikkimese society; so much so, that it is mandatory to reserve one seat for the 'Sangha' in the State legislative assembly. The obligation of an average Sikkimese to the 'Sangha' is manifested in every aspect of his life from birth to death. Non-Buddhists of Sikkim are often found to be guided by the lamaistic rituals in their hours of distress. An average Sikkimese, irrespective of his religious affiliation, pays homage to the Buddhist deities as well as high priests.

If the ethnic scene of Sikkim is dominated by the Nepalis, the religious scene is undoubtedly dominated by the Bhutias, who still envisage themselves as superiors, standing higher than the rest of the society. The overwhelming figurative increase in the number of Nepali population and general acceptance of Nepali language as the 'lingua franca' of Sikkim is counter-balanced by the supremancy of the Sangha and the tribal elites over the Sikkimese society.

The Stir

The motion against the project was first surfaced in 1994 when the Denzong Lhaday Yangki Chokchen (Association of Buddhist Monks of Sikkim) officially raised objection to the project by submitting a memorandum to the then Chief Minister of Sikkim, S. Limboo, during one of his visit to West Sikkim. The memorandum demanded immediate removal of the project from the Yuksam area. In the same year, several tribal associations took part in the stri and filed law suit in the High Court of Sikkim. In 1995, a group of Bhutia activists under the banner of 'Concerned Citizens of Sikkim (CCS)[14], appeared in the scene with the demand of immediate scrapping of the project.

According to this group, "the environmental and socio-cultural costs of the project would be too high a price to pay for its hypothetical economic benefits".[15] Following the example of Medha Patkar on the Narmada issue, one of the leading members of the CCS, S. Denzongpa, began hunger strike. However after 28 days, the hunger strike was called off following the state government's decision to constitute a high level committee to look into their grievances. The CCS issued a press statement stressing... "The project will have to be reviewed in light of it being environmentally unsound, economically disastrous and imposed on the public as a 'fait accompli'. All this was due to the conspiracy of corrupted officials and well-connected commission agents and contractors whose greed proved no match for the negative environmental, historical, religious and socio-economic Consequences the project will cause..."[16]

Meanwhile the monks of Sikkim submitted another memorandum[17] to the Chief Minister stating,

"Hon'ble Sir,

18 June, 1995

The Buddhists of Sikkim hereby wish to express their objections regarding the proposed Rathong Chu hydel project... the project will destroy the hearth of the most sacred area of Sikkim which has enormous historical and religious significance for our people. This issue is not a question of mere 'sentiments' or 'emotion' hindering progress and development as some non-Buddhists might have insinuated... It is unethnical to sacrifice the basic foundation of a vulnerable minority such as its history, culture, religion and environment for the benefit of others. By now, the authorities should have learned from the rise of fundamentalism in the rest of the country, that they should be more careful and avoid attacking people from a religious angle".

> ".. we would like to submit that from a religious point of view the proposed project is undesirable as it is located at the heart of our place of worship, that it will defile the most sacred region of Sikkim and that the Rathong Chu hydel project goes against the religious sentiments of the people.
>
> "Thus, we request you Hon'ble Sir to intervene in the implementation of the Rathong Chu hydel project by stopping all work at the site"....

Buddhist Monks of Sikkim

In another memorandum, dated 29.7.95, the Sangha of Sikkim wrote,

> "We, the monks of Sikkim hereby demand that the Rathong Chu hydro-electric project located at Yuksam be stopped immediately. We oppose the project because it will have severe negative consequences for the most sacred area of Sikkim...
>
> "We oppose the Rathong Chu project because it will... defile the whole area which is sacred to us.... It is offensive to see outside labourers washing meat next to Chorten, dumping excavated mud around the Kathok Lake, and to witness that the human waste of these labourers are dumped into the sacred Rathong Chu and will end up in the Bum Chu of Tashiding"...
>
> "The Nay sol text is to the Sikkimese Buddhists what the Ramayana is to the Hindus, the Koran to the Muslims and the Bible to the Christians. If the Rathong Chu-hydro-electric project is not stopped and abandoned, we the lamas of Sikkim are ready to burn our Nay sol texts as its meaning and purpose will be lost...

"This time we hope that our voice will be loud enough so that it may not be ignored".[18] In retaliation, the lamas of Sikkim took to the street. Their year long agitation culminated in thronging the High Court Premises as well as the State Secretariat. However, on the whole, the mob was not violent, no casualty was reported either.

Subsequently, another writ petition was submitted to the High Court by the president and members of the CCS in August 1995.[19] This time, to strengthen the cause, the petitioners mooted the question of conservation of environment along with conservation of their heritage.

All the allegations against the project ultimately proved to be futile, because, on enquiry by different high level committees, it was proved that, the construction of the Rathong Chu Hydro-Electric Project does not involve mass deforestation, nor does it involve permanent evacuation of any family. The colony is situated in a barren land. It does not interfere with any religious or cultural or historical monuments.[20] The nearest religious-cum-historical monument, the 'Norbu Gang' lies at least two kilometers away and at a much higher altitude. The Dubdi monastery, which was repeatedly cited by the petitioners to be in danger, is several kilometers away from the project site.

Consequently, both the writ petitions were rejected by the court on the ground that they were devoid of merit.[21] Following the verdict, the agitation subsided but the undercurrent still persists.

Conclusion

Two things are clear from the protestations. Just like in some other part of the country, the people of Sikkim are averse to part with their religious sentiments even at the cost of development and prosperity. The water of the Rathong Chu, which is believed to be so sacred, is being used day to day by the local inhabitants for centuries just the way the water of the holy Ganga is being used by people since time immemorial, although on different scales. If daily use by the local inhabitants does not alter the purity of water then it is irrational to think that the construction workers of the project will render it impure by domestic use.

Secondly, and more importantly, the ethnic Bhutia-Lepcha population of Buddhist belief are not so much worried about water pollution or environmental degradation as they are about immigration of outside labourers in a land which is a strong Bhutia enclave. The worry of the local inhabitants is not fully unfounded. Because, there are precedences of similar mass immigration in the history of sikkim. Way back in the 1880's, mass immigration took place in Sikkim when the copper mines of East Sikkim were opened for the first time.

Since superstition prevented the indigenous Sikkimese from mining activities, they hired mine workers from the neighbouring Nepal,[22] who, within decades, outnumbered the Bhutia-Lepchas. Another mass immigration, mainly or white-collared workers, took place in the wake of Sikkim's merger with India in 1975 and thereafter. By 1980, the tertiary sector of the state was visibly dominated by the plainsmen coming from the adjoining states of West Bengal, Bihar as well as from distant Punjab, Rajasthan, Haryana and even as far as Kerala. The Bhutia-Lepchas, who have already become ethnic minority in Sikkim are afraid of further submergence by people of alien culture.

Acknowledgement

I am grateful to Hon'ble Justice M. Sengupta, High Court of Sikkim, and Mr. T. Bhutia, Engineer-in-charge, RHEP, Yuksam for providing me with necessary data and informations. I am indebted to Mr. T. Adhikary, field supervisior, RHEP for helping me in my fieldwork at Yuksam. I am also thankful to the Director, Centre for Himalayan Studies, North Bengal University for sponsoring my fieldwork.

References

1. WAPCOS : Rathong Chu Hydro-electric Project, Sikkim. Detailed Project Report, Vol. II Chapter 3,

Prepared by water and Power Consultancy Services (India) Limited (WAPCOS). July, 1989.

2. Sikkim Express : "Water Resources That Flow From Sikkim: An Economist's' View" Sikkim Express; 1st July-7th July, 1995, Gangtok.
3. *Ibid.*
4. *Op. cit.*, 1.
5. *Op.cit.*, 1.
6. *Op.cit.*, 1.
7. *Op.cit.*, 1.
8. Manisha Aryal : "The vocabulary of Protest", Himal, July-Aug, 1995, Vol. 8, No. 4, pp. 11-21.
9. Sikkim Express : "Rathong Chu Hydro-Electric Project Domestic Resource Mobilisation Effort of Sikkim; Sikkim Express: 1st July-7th July, 1995. Gangtok.
10. L. Tiwari : "Affidavit-in-opposition in the High Court of Sikkim", 1.9.95.
11. Writ Petition No. 10 of 1994, in the High Court of Sikkim.
12. R.N. Dokhampa : "Origin of the Bum Chu of Dakkar Tashiding". Bulletin of Tibetology. SRIT, New Series, No. 3, 1992, Gangtok
13. *Ibid.*
14. Ludwig Schaefer : "A Sikkim Awakening" p.10, Himal, Sept-Oct. 1995, Kathmandu, Nepal.
15. *Ibid.*
16. Sikkim Express : "Hunger strike on Rathong Chu called off", Sikkim Express, July 15th to 21st, 1995, Gangtok.
17. Memorandum submitted by the Association of Buddhist Monks of Sikkim to the Chief Minister of Sikkim dated 18th June, 1995.

18. Memorandum submitted by the Sangha of Sikkim, Dhotapo, Deorali to the Chief Minister of Sikkim dated 29th July, 1995.

19. Wirt Petition No. 25 of 1995, in the High Court of Sikkim, Gangtok.

20. Sikkim Express : "Compensatory Afforestation and non-interference of Religious Place", *Sikkim Express*, 1st July-7th July, 1995, Gangtok.

21. Sikkim Express : "Sikkim High Court Quash Writ Petition on Rathong Chu"; *Sikkim Express*, 22nd Dec. 1995.

22. Risley, H.H. : *Gazetteer of Sikkim*, Calcutta, Bengal Secretariat Press, 1894.

11

Inter-Ethnic Relations in Sikkim in Historical Perspective

Objective and Plan of Study

Sikkim, on of the smallest states of India, has been inhabited by a number of communities, the most important of which are the Lepchas, the Bhutiyas and the Nepalis. Of these, the Bhutiyas, who represented te ruling class, headed by the Chogyal, came from Tibet in the 17th century and established their supremacy over the Lepchas. They, in their turn, were later, outnumbered by the Nepalis, who immigrated en masse to Sikkim in the 19th Century. Their social and political importance, however, continued unabated till 1975, the year, when Sikkim finally merged with India to become its 22nd State. After this, the floodgate of immigration was opened for the plainsmen from different part of India as well as for other Nepalis from the neighbourhood. All this added to the socio-economic complexities of the state and shaped the interethnic relations.

In this study we have tried to explore, how this relation in Sikkim has evolved over a long period of nearly 400 years an have tried to identify the major phases an the issues that have gone to shape it.

The focus is on the Lepchas, the Bhutiyas and the Nepalis in the pre and post merger periods. The paper is divided into the following sections:

1. Introduction: A demographic profile of the state.
2. The autochthonous versus the Bhutiyas.
3. The Nepalis immigration.
4. The polarization along ethnic lines; A development of the 1950s.
5. The major issues since 1975.
6. Conclusion.

Lepchas, Limbus, Bhutiyas and Nepalis: The 4 major communities.

H. H. Risley, in his article in the Gazettes of Sikkim, identified three main stocks in Sikkim.[1]

1. Rongs or the Lepchas as the Nepalis call them.
2. Khampa or Kham-ba, the immigrants from the Tibetan province of Kham, commonly called Bhutiya.
3. The Limbus of Lhasa Gotra who are believed to have migrated to Sikkim from Shigatsi, Penam, Norpu, Khyongtse, Samdubling and Gyantse-all places in the Tibetan province of Tsang, South of Tsanpo.

A fourth stock that has to be added to his classification is that of the Nepalis, encompassing several tribes who immigrated from 19th century onwards. Perhaps it was their late arrival that was the reason why Risley left them out of his classification of the people of Sikkim.

In the absence of any written script of the languages of the indigenous people in the people in the period to the arrival of the Tibetans, one has to depend on their mythologies and oral traditions for a reconstruction of their history. These oral traditions, obscure and contradictory at times, do not always lead one to a definite conclusion. Among the oldest and perhaps the aboriginal inhabitants of Sikkim were the Lepchas, who, in all probability were accompanied by some Libus and a few Mangars.

The Limbu Tradition

The Limbu tradition does not maintain distinction among these three tribes but includes them among the Kirati stock ruled by their tribal chiefs. These Lepchas had a simple material culture-a subsistence life of hunting and collecting as well as a primitive form of slash and burn type of rotational cultivation of rice, maize and millet. Their religion was basically animistic Bon and their priests a combination of holy specialists, magicians and medicine men. Risley mentioned 39 Lepcha families still existing particularly along the banks of the Talung river.[2] Col. Mainwaring reports of a Lepcha Pano (king) Turve ruling around AD 1425, hwne a Bhutia patriarch-khye Bumsa, an ancestor of Phuntsog Namgyal came to Sikkim form Kham in Tibet. He came across a Lepcha Chief-The Kong Tek. Thelast Lepcha Pano was born in 1686 and ruled for 25 years.[3]

The origin of the Limbus is veiled in obscurity. There are three/septs among them-one belonging to the Kasi Gotra, indicating that they cam from Benaras and the Lhasa Gotra as they are believed to have migrated from Tibet. The third is known as Bhuiputha. While those belonging to Lhasagotra follow Lamaistic Buddhism, those belonging to the other two septs are animistic and to some extent Hinduised.[4] Risley narrates a Limbus tradition taken from an old Limbu manuscript that describes how one of the two branches of the descendants of suwang be-ba (progenitor of not only the Limbus but the entire human race) appeared at Kasi, while the other traveled East, into the hills and in subsequent period came to be known as Munapemba or late comers. They came ot be known as the Lhaasagotra for having come form the direction of East Tibet.[5] De. Waddell explains that the name Limbu has been given to them by the Nepalis. They call themselves Yakthumba or Yakherds while the Lepchas and the Bhutiyas call them Tsong meaning a merchant.

The Limbus were the chief cattle merchants and butchers in Sikkim.[6] The Bhutiya immigration took place in phases and Risley has identified the original families and the sequence of their arrival. The earliest immigrants represented 14 main families of Tibeten origin. Of these, 6 were descendants of Khye Bumsar and 8 were of other Khampas now known as Lbeps-m TshanbGyad or the tribe of 8 respectable names. In addition, there were 8 Rui Chhung or little families who can to Sikkim at various times since the establishment of the monarchy. After all these came the Chumbipas, immigrants from Kham in Tibet and Hah in Bhutan to the lower reaches of the Monchu near Chumbi. They were subdivided into groups.

Colonisation

In the North, the Lachen Lachung valleys were colonized by other immigrants from Hah and Paro in Bhutan, now known as the Lhopon Lhondub. Finally, Risley identifies another category in which he gives name of 12 other familes.[7] These Bhutiyas, representing the Tibeto Burman stock, introduced Tibetan culture, Lamaistic Buddhism and a combination of pastoralism and semi settled agriculture in Sikkim. On the other hand, the Nepalis contributed the Indo Aryan languages, Hinduism and settled agricultural practices. The term Nepalis, however, is a very broad and general term, encompassing within itself, a number of hill tribes as well as Hindu castes.

The process of identity formation and acculturation is not complete for some of these ethnic groups as yet. Some ethnic groups, like Rais, Mangars and Gurungs, originally considered tribe have now become a part of Hindu caste system. Again, some of them are demanding separate tribal status, eg. The Limbus and the Tamangs, on ground that they are closer to the tribes than to the Hindu Nepalis in matters of rituals and practices.

The Ethnic Language

As many of the ethnic groups, originally considered tribe did not have their written script (the Limbus are an exception) their language came to be considered a dialect of the Nepali language. Since Nepali is the Lingua franca of all hill people in this region, all these groups of people, speaking the Nepalis language are now known by the general term Nepali. Religion wise also there is diversity among them. The Tamangs and a section of the Limbus were originally Budhists, though now they have adopted some of the Hindu rituals and practices.

Sociologically, they are divided into three sub-cultural stocks. The Kiratis, the Newaris and the Gurkhas.

1. The Kiratis includes the Limbus, Rais, Magars, Gurungs, Tamangs and a host of marginal tribal stocks who can again be broadly classified into three groups on the basis of their absorption into the larger civilizations:

   (a) The animists like the Gurungs, Magaras, apart of Limbus. b) The Hinduised Rais and the Lamaist *Tsongs*, Tamangs and Sherpas.

2. The Gurkhas included the Khas, the Thakuris and the Chhetris who came to settle down in Sikkim in the middle of the 18th Century after the emergence of Gurkha power in the form of Prithvinarayan Shah in Nepal.

3. The Newars entered Sikkim in the 18th-19th Centuries after the liquidation of Newari power in the Kathmandu Valley by the Gurkhas.

With them cam ea large number of service castes like the Kame (Smiths). Damai (tailors) and Karke (Shoe makers) The Newaris are, by and large Hindus with the highest socio-economic positions given to the Pradhans followed by the ritual leaders-the Brahmins.[8]

Population Figures

The earliest information that we have of population figures of Sikkim is the 1891 census report. It ran as follows:[9]

Population Sikkim

Total Population	-	30458
Lepcha	-	5762
Bhutiya	-	4894
Limbu	-	3356
Gurung	-	2921
Murmi	-	2867
Rai, Jimdar etc.	-	2020
Khambu	-	1963
Kami	-	1670
Brahman	-	1414
Mangar	-	901
Chhetri	-	829
Newar	-	727
Slaves	-	326
Dorzi	-	287
Misc. including troops	-	521

The subsequent population figures give an idea about the pace at which the Nepali population increased in Sikkim. Writing in 1909, J.C. White gve the number of Lepchas inhabiting in Sikkim at about 6000, that of Bhutiyas, a little over 6000 while the number of Nepalis were 50,000.[10]

As per the 1931 census, out of a total population of 109808, 13060 were Lepchas, 11955, Bhutiyas and the rest mainly Nepalis.[11] In the census reports of 1951 and 1961 we find categorisatio of population on the basis of mother tongue while in 1981, it was the language spoken in the household. Though the basis of categorization was more or less the same in 1971 there are some discrepancies in the two reports. In 1971, there was no separate group for Limbus or the Sherpas which was mentioned separately for the first time in 1981.

Secondly, there is the use of the thre terms Tibetans, Bhutiyas and Sikkim Bhutiyas to indicate the people of Tibetn origin. While Tibetan probably indicates the Tibetan refugees in Sikkim, the difference between the Bhutiyas and the Sikkim Bhutiyas is not clearly spelt cut. This distinction was done away with in the 1981 census.

The following table give the population figure of Sikkim on the basis of mother tongue or language spoken in the household as per the 1971 and 1981 census reports.

Table 11.1 : Population of Sikkim on the basis of mother tongue (1971) or language spoken in the households (1981)

	1971	*1981*
Bhutiya	8064 persons	21548 persons
Tibetan	4607 persons	4149 persons
Lepcha	2231 persons	22391 persons
Sikkim/Bhutiya	10891 persons	-
Nepali	134235 persons	192891 persons
Limbu	—	17922 persons
Sherpa	—	10726 persons

Source: Sikkim census, 1971 and 1981.

The total number of population of Sikkim as per 1991census is 406457. Of these 22.36% are reported to be tribals. District wise, the per centage of the tribals is as follows :

North	-	55.38
East	-	21.09
South	-	16.91
West	-	19.66

Within the tribals are included 1. Bhutiya (including Chumbipa, Dopthapa, Dukpa, Kagatay, Shera, Tibetan, Tromopa Yolmo) and 2. Lepcha.

Detailed breakup of population figures on the basis of mother tongue or language spoken at home is not yet published.

The Major Ethnic Groups

The three major groups of Nepalis, Bhutiyas and Lepchas, whether judged objectively or subjectively, are generally accepted as three ethnic groups. Each is different from the other by historic antecedents, geographical centres, cultural emblems, etc., which are accepted as components of objective view of ethnicity. Coupled with this is their self ascribed awareness of distinctiveness and a feeling of belonging to the "group", which is recognized as a group by others as well. This is the major component of subjective view of ethnicity. Therefore, it is a self defined as well as other recognized status.

Since ethnicity also involves competition for resources of a certain territory, to the claim of ethnicity by these three ethnic groups of Sikkim is closely linked the question of equal rights and opportunities in politics, economy and society. In fact, it has always been this question of equal rights that has shaped the ethnic relations in Sikkim. The administrative policy of the Chogyals of Sikkim had gone a long way in shaping this relation in pre merger period. During this period, the attempt was to maintain the Lepcha Bhutiya supremacy by giving special rights and privileges to them. This, in its turn gave rise to a feeling of discontent in the minds of the Nepalis who felt that despite being greater in number they were inferior in status to the Bhutiya-Lepchas.

Political Scene

The political scene has changed after the integration when democratic government was installed in Sikkim. Yet, the legacy of that system is, to some extent still determining the main political issues and inter-ethnic relations. Now it is turn of the Bhutiya Lepchas to suffer from a feeling of deprivation. Even though they are still enjoying certain privileges by virtue of being the original inhabitants and ex-rulers of the state, they feel they are losing their old status and importance in the new democratic set up. This has brought them together to fight for and maintain some of their old rights. In the subsequent sections we have described the process and the main issues.

The Autochthonous versus the Bhutiyas

As already stated, it is a tough task to reconstruct the history of Sikkim before the Tibetan migration of the 17th century due to the lack of any written script of the original inhabitants in this period and the consequent absence of written account of the course of events. All that we have are a few stray references of an abstract character scattered in the vast body of Tibetan literature where secular accounts and narrations of religious events of a mythical nature are mixed up in such a way that it is not always easy to segregate the two. The accounts beome detailed and authentic only from 1642 onwards, when Phuntsog Namgyal was consecrated by Tibetan lamas as the first king of Sikkim. From the Reference that we have of the period prior to this, it appears that the Tibetans used to visit Sikkim more or less regularly because they considered the ladn extremely sacred. The relation between them and the inhabitants used to be cordial.

The Bhutiyas versus the Lepchas

The earliest reference that we find of interaction between the two communities of Bhutiyas and Lepchas involves one The King Tek–a great Lepcha patriarch and wizard to whom even the Tibetan came for help. Gyad Lbum Sag, grandfather of Guru Tashi and great grandfather of Phuntsog Namgyal even got his there sons blessed by him.

This created a bond of friendship between the two communities which was 'cemented by rituals and ceremonies, sacrifice of animals and invocation of spirits'.[12] The Bhutiya kings tried to retain this good relationship with the Lepchas by inducting them into the fold of the administrative set up. Huntsoog Nemgyal selected 12 Lepcha Jongpens from among the superior families and appointed them as kazis who thus came to form a section of aristocracy along with the Bhutiyas. Proclamations were made, promising the recognition and emoluments to those who distinguished themselves by loyal and faithful services. Posts of ministers and prime minister *(Chag–mzod)* would be conferred on them. On the other hand those who did not serve well would be classed among the common people and required to contribute much services as were required by the Maharaja.[13]

The Trustworthy Lepchas

The Bhutiya sources claim that from the very beginning, the Lepchas flocked to the service of the new king. The common Lepchas, who roved to be trust worthy were appointed in the household establishment of the king, while others were entrusted to other state duties. Finally, those who were not so much in the confidence or favour were employed to carry goods and come to be called *Tsong Skyal Monpas.* Some of them were even employed as assassins, or labourers in construction of buildings and forced to perform other manual work. They were also expected to pay the tax of Yar-jal, in form of crops, grains etc.[14] The Lepcha commoner seem to have been overshadowed by the culturally more dominant Bhutiayas. Driven to the woods, they led the life hunters and collectors of wild roots.

The Bhutiya king also used matrimony as an instrument of cementing their relation with the Lepchas. King Tansung Namgyal also carried out an extra marital relationship with the wife of Lepacha headman Tasha Apong. The son born out of this relationship was Gugthing Arup, who was later appointed head of the Lepchas and subsequently become the *Chagzod.*[15] King Chagdor Namgyal also studied the Lepcha language and invented a new script for it. His own rituals, in fact, were written in it.[16] But underlying this apparent cordial relationship that the history written by the Bhutia kings themselves like to highlight, was an undercurrent of tension that is brought out by a careful of the course of events.

The *History of Tibet* records several instance when the Lepcha 'sercerers' tried to cast spell over the Bhutia king. For example, there was an attempt to raise a general insurrection by an 'eveil spirit', which had brought in its train ill luck and every untoward misfortune to the Sikkim State. In general and to the Raja and his connections oin particular. First he had entered the Rabdentse palace from where he tried to diminish the prosperity of the Sikkim state and later took refuge in Rinchenpong where he conspird to raise a general insurrection. Having failed in this too, he went back to Pasak to collect a body of 14 males and female 'Bijuas' and Bijuanis of the Lepcah race called Bonbons. He employed these sercerers to work mischief by means of the black art but this product no effect on the Raja...[17] The person thus possessed was finally enslaved and engaged in fetching water, splitting fuel etc.[18]

Again, there was the incident of 5 Lepchas, who, pretending to be the incarnation of great Lepcha wizard Guru Teshe Thing tried to bring the king under their influence. As per the account recorded in the *History of Sikkim*, thee Lepchas even succeeded in convincing the king of their supernatural power and had almost won him over. The Buddhist Lamas however, were not to be taken in so easily. Finally they prevailed over the king and roved that thee 5 Lepchas did not have any supernatural power at all. At the end, the trapas of the Pemioche monastery stoned them to death.[19]

Incarnation

During the reign of the same king, another Lepcha subject, Tisho Bidur by name, declared himself to be the incarnation of Guru Rimpoche. He too exhibited some miraculous powers and even collected round him a large number of followers. He too raised a rebellion against the king and with the help of the Mangar Raja, stopped payment of revenue to the king. This revolt also was crushed and the rebels killed.[20]

Now these events have to be interpreted in the historical context of the period. This was the period of the spread of Buddhism in Sikkim. New monasteries were being built and Lepchas converted to Buddhism. The king himself was person of religious disposition. The attempt on part of the Lepchas to kill the king may be considered in this context. If the king was won over by them, the Lepcha religious Gurus would find it easier to regain their control on the masses.

However, the Buddhist Bhutiyas were careful enough not to alienate the masses too much. Hence, after the Raja's death, donations were distributed among the Lepcha Tassas and the Raiyats along with the Bhutiyas. Finally the Bhutiyas succeeded in extending their sway over the Lepchas most of whom gradually converted to Tibetan Buddhism. This was made possible by the increasing number of Gompas or monasteries whose one of the primary functions was to convert pagan Lepcha to Buddhims.

This is the view held by Chie Nekane who argues that this is how the closed local communities opened to the outside world and a primitive society (that of the Lepcha) changed into a peasant society under a centralized political system connected with the greater world of Tibetan civilization.[21] Their native belief of Tibetan Buddhism were closely interwoven in their culture and produced one of the distinctivo frontier cultures within the Tibetan cultural area.[22]

Intermarriages

The intermarriage between the Bhutiyas and the Lepchas of the upper strata of society acted as a second agent of integration. From such intermarriages developed a section of Lepcha aristocracy. For much of the time the Lepcha aristocrats held their own but gradually they became cut off from the Lepcha masses and began to identify themselves with the Bhutiya aristocrats.

The Bhutiyas versus the Limbus

As to the Limbus, the relation between them and the Bhutiya king's was initially cordial. As in this treatment towards the Lepchas, the king Tansung Namgyal had tried to establish matrimonial relation with the Limbus as well. He himself married the daughter of a Limbu chief who ruled in the Arun Valley in the West of Sikkim. A.C. Sinha points out that out of 12 rulers, 7 married Limbus ladies from Limbuan.[23] This practice was enthusiastically followed by the ministers as well.[24] But this cordial relation was not destined to last long. There are references to revolts raised by the Limbus.

In 1741 for example, one Limbu, Sri Junga Devasi revived the Karati Mundhum religion. The Bhutiya Tachang Lama of Pemiengchi monastery put him under arrest and short him dead. There was a second case of Limbu revolt in 1752 which was crushed by the Durbar.[25] Finally, they migrated anmasse to Limbuana. According to the *History of Sikkim*, their discontent was a result of their being overworked at the construction works of the pathway between Dechen Ling and Rabdentse palace as well as the monastery in Chen Ling Palace.[26]

The Bhutiyas versus the Mangars

Among the autochthonous people of Sikkim, there were Mangars, through very little is known about them in this period. H.H. Risly in the *Gazetteer of Sikkim* citas one incident in course of the reign of Phuntsog Namgyal which, in all probability, has a bearing on the Mangars. It is said that with the aid of Lha-tsan Lama, he overcame one Shintu Satichen or Mangal Gyelpa. Though he is generally considered a Lepcha, Risley contends that the name sounds more like a Mangar. This tribe, Risley maintains, occupied the valleys to the south of Knachenjungha Everest range. The chief was said to have disappeared without leaving a trace.[27] The Mangars too were overworked by the Bhutiyas which alienated them from the rulers. Around 1750 the Mangar chief of Sikkim Terai outraged by the discourtesy shown to him by the Sikkim Durbar, even persuaded the Deb Raja of Bhutan to attack Sikkim.[28]

The arrival of the Bhutiyas in a way brought about some basic changes in the demographic profile of Sikkim. The Bhutiya rulers in order to establish their authority played the traditional methods of matrimonial alliances and concepts of ritualistic bons. The *Lhomentsongsum*, an association of the trio of *Lhopas* or Bhutiyas, *Monpas* or Lepchas and *Tsongs* or Limbus was formed by Phuntsog Namgyal out of the concept that all the three tribes belonged to one family. The king, representing the Bhutiyas was the father, the Lepchas the mother and the Limbus were the sons of this family.[29]

Carrot and Stick

In addition to this, the method used by the Bhutiya rulers to allure the other tribes to submission to their authority was one of carrot and stick. Those who willingly accepted their domination were offered high posts while force of some kind was used against those who did not. The declared policy of the Bhutiya king was not to allow anyone to 'remain masterless as before'.[30] Yes, they failed to win over minor tribes like the Limbus and the Mangars by use of theise methods. Incidents of revolts and expressions of discontent against the Bhutiyas continued sporadically and finally, unable to maintain their independence against the dominant Bhutiyas, many of them migrated to neighbouring countries.

Barring a few cases of revolt against the Bhutiyas at the beginning, the history of Lepchas in this period was one of gradual assimilation with the Bhutiyas. As a prize of their collaboration, they gained not only material benefits like high posts and landed property but also the prestige of being equated with the ruling Bhutiyas as the Buddhist tribes, representing the ruling class. The Bhutiyas and the Lepchas were to oppose the Nepali immigration and settlement in the 19th century together.

The Nepali Immigration

Nepali immigration to Sikkim was a phenomenon of the 19th Century, a result of direct encouragement from the British. In contrast to the Bhutiyas and the Lepchas, the hardworking Nepalis made a very good peasant in the peace time just as they made officient warrior at the time of war. It was a calculated policy on part of the British to settle these people in the sparsely populated stretch of land of Sikkim. About these people J.C. White writes.[31]

> "They [Pharias=Nepalis] have migrated from the neighbouring densely populated state of Nepal

and are slowly but surely pushing their way eastward. They are almost all Hindus by religion with innumerable castes, the few exceptions, being the tribes coming from the North-East of Nepal, who still profess Buddhism. They are on the whole a steady, industrious and thrifty people, very pushing and eager to take up new employments, they make excellent settlers, pay their rent regularly and give no trouble in way. But they require a strong hand over them and some of the castes are most litigious and quarrelsome. Many of their headmen are excellent managers, thoroughly to be trusted and will carry out anything they undertake to do to the best of their ability. In more than one case of I have known Nepalese settlers in Sikkim, by dint of hard work and perseverance, rise to important positions which they have successfully filled, in marked contrast to the Lepchas, whose indolent temperament always act as a deterrent and causes them to be outdistanced by more energetic races."

Religious Confrontation

By the time the Nepalis started arriving in Sikkim, the acculturation between the Bhutiyas and the Lepchas had reached such a high degree hat these two ethnic groups, professing Buddhist religion confronted the Hindu new comers as one integrated Buddhist population. Te differences between the Buddhist and the Hindu values, ideas and life styles were great and assimilation of these two groups was never as high as had been the case of the Lepchas and the Bhutiyas.

In fact, the early years of Nepali immigration were full of strife on account of the opposition of th Bhutiyas. In this, the Bhutiyas received direct encouragement from the Chogyal. From the very beginning, the Raja's Council was divided into two groups on this issue. Majority of the members, including the king Tsugphus Namgyal were opposed tooth and nail to Nepali immigration.

His dislike of te Nepalis emanated, to some extent, from his personal experience. This was a result of the repeated raids of the Bhutanis and the Gurkhas into Sikkimese territories and the miseries to which his father had been subjected to as a result of these inroads.[32] In fact, the king Tsugphu Namgyal always engendered such a feeling of bitter enmity for the Bhutanese and the Gurkhas that he had always expressed his feeling by enjoying upon the ministers, Lamas and Kazis as his irrevocable will that not even a single person belonging to thee states should be allowed to settle down in Sikkim.[33] This order was reported again and again.

Majority of the member of the Raja's Council shared this feeling of the king and opposed any move to allow the Gurkhas to settle in Sikkim. Despite such vehement opposition from the king and his Council, the Nepalis did manage to get the permission to settle in Sikkim. This is thus described in the *History of Sikkim*.

"In the year 1875, in direct defiance of the Royal order, Chebu Aden Lama allowed the Paharis (Gorkhas) to settle n the land of Chakoong Rishi and Ramon rivers. His example was soon followed by the old Lasso Kazi who allowed Paharis to settle in Katam and namchi and without so much as asking permission from the Maharaja he opened and began to work a limestone quarry and make lime. The contagion of disobedience spread the Phodong Lama and the Khangsa Dewan also allowed Paharis to settle in Tchadam."[34]

The next king Sidkoong Namgyal, however, was more practical and in order to ingratiate himself with the British authorities, granted the first lease to the Gorkha settlers in 1867, viz. to the two Newar–the trader Laxmi Prasad being one of them.

This period of leniency was short-lived as the policy was once again reversed by Thutob Namgyal, the half brother of Sidkeong Namgyal immediately after his accession. He took an intense dislike to the Nepalis from his father.

The Council became divided into two factions–pro and anti immigrants–and the matter reached such a height that the king was left with no other option but to seek the advice from the Governor General of Bengal, Sir Ashley Eden.

The British Tilt

Though sympathetic to the king, he gave his final verdict in favour of the Gorkhas. To quote from his letter. "The lands of Sikkim belong to the people of Sikkim and to no one else. If the wasteland and unihabited places are to be given to the Newars, yet the headmen should be Sikkimites. On no account should any native be deprived of their lands to be given to strangers and aliens. Although your Highness may say that you had consulted Sir George Campbell and Sir Richard Temple upon this question of allowing Newari settlers to populate the country and that they had expressed consent, yet, I am quite sure that they never thought that native Kazis, Kyomees, and headmen should be deprived of their patrimony for the benefit of such utter strangers as the Newars are..."[35] He advised the king to take personal interest in the claims and titles of the lands belonging to each and everyone of Sikkim subjects and not to be guided by the reports and statements of one or two ministers.

Finally, a personal meeting between the Chogyal and the Governor General was arranged in Kalimpong. In it Sir Ashley Eden repeated his previous suggestion. "It would be better if Gorkhas were not all all allowed to settle in Sikkim. If it be necessary to populate the water tracts of Sikkim by filling them with Gorkha settlers, yet they should never be allowed to become headman or obtain any power of officers..."[36]

Strife

The bitter strife that marked the early days of Nepali immigration gradually subsided with the passage of time, the Nepali peasant began to live in amity with his Bhutiya and Lepcha counterpart albeit as two separate ethnic groups. There were not many instances of confrontation between them at a lower level. Inequality persisted at a higher level where some Bhutiya landlords still discriminated against the Nepalis and the laws of the land were also framed in favour of the Bhutiya and the Lepchas.

The Polarization Among Ethnic Lines: A Development of the 50s

The polarization of politics along ethnic lines was a development of the 1950s. It was in this period that Sikkim's relations with Independent India was formalized and political parties were formed by educated elites who were inspired by the freedom movement in India. In this period of turmoil and political transformation in the state, the inter ethnic relations took a turn towards communalism. The discontent of the Nepalis with the Chogyaol's pro-Bhutiya-Lepcha policies (discussed later in this and the net section) began to be expressed openly through the various agendas and demands of the political parties.

The Durbar in its turn, patronized some other parties to counteract these demands of democratization. Such accusations and acrimonies brought out the mutual antagenism of the Bhutiya Lepchas and the Nepalis to the open a feeling that though always there, had been lying dormant at the social level for so long. The first two political parties that moulded the course of consequent course of political event in Sikkim were the Sikkim State Congress and the National Party.[37] The Sikkim State Congress stood for democratic ideals like abolition of landlordism and democratization of administrative machinery. They even supported the idea of accession of Sikkim with India.

The National Party

The National Party, patronized by the Durbar, was basically traditionalist and emerged in order to counteract the demand raised by Sikkim State Congress. It described the demands of democratization of administration and abolition of landlordism as a luxury which a small state like Sikkim could not afford. It drew its support mainly from the landholding and monastic elites. However, following extensive agitations by the Sikkim State Congress the Durbar had to give in to some extent. On the advice of the Indian Political Officer, the Chogyal agreed to form a popular ministry under the Chief Ministership of Tashi Tshering, the President of Sikkim State Congress. A 5 member Council of Ministers 3 of whom were Congress and 2 Durbar nominees took oath of office on May 9 1949.[38]

Signing of Treaty

It was at this time a treaty was signed between Sikkim and Independent India in December 1950. The Sikkim State Congress felt let down by the features of the treaty as they found that the wishes of the people were ignored. The establishment of responsible Government, peoples' participation in the administration etc. were nowhere mentioned in the treaty. They interpreted it as an attempt on pat of the Government of India to placate the Chogyal. The Government of India, they alleged, had come to a secret understanding with him cut of a fear of Chinese presence on the border of Sikkim.[39] The party's agitation for democratization continued. It was in this context of political unrest that the egap between the Bhutiya Lepchas and the Nepalis widened.

The immediate backdrop was provided by the Panchayat election of December 1950. In the 1940s the Sikkim State Congress had demanded the formation of Panchayats and transfer of powers from landlords to village Panchayats. As per the notification of 12 June 1948 Panchayat tribunal consisting of a landlord and 4 villagers of an estate was formed, the selection methods were also laid down. Though the provision, under the notification sought to herald participating democracy, the attempt became virtually abortive in view of confrontation between the Durbar and the people led by the Sikkim State Congress when election was scheduled to be held in December 1950. The National Party boycotted it on the ground that the interest of the indigenous population had not been safe guarded. By indigenous population it meant the Bhutiyas and the Lechas because these people were already there when the Nepalis poured in.

The National party representing forces and backed by the Durbar, now openly sided with the Bhutiyas and the Lepchas. Their boycott of the Panchayat election was played up as the expression of the dissatisfaction of the indigenous population with the system of Panchayat elections. They also made the propaganda and succeeded in convincing the masses to a large extent that the State Congress, though headed by a Bhutiya, actually sought to dominate the indigenous Bhutiyas and the Lepchas with the help of the Nepalis.[40] As a result of this, the State Congress began to lose the support of an increasing number of Bhutiyas and Lepchas. Even though the latter talked in terms of unity of the Sikkim people and ideals of democracy it, by virtue of drawing its support mainly from the Nepali population, came to be regarded as a party representing Nepali interest alone.

Communalization

This trend of communalization of politics ultimately culminated into the introduction of the party system. Out of 12 elected members of the Council 6 seats were decided to be kept reserved for Sikkim Bhutiyas or Lepchas and the remaining 6 for Nepalis. The Maharaja was to appoint 5 nominees.[41] There were two Royal Proclamations of January 1952 and March 1953laying down the rules regarding formation of the State Council an the Executive Council, mode of elections, powers and functions. The mode of election to the Council was rather complicated. Each of the Nepali seats had to be returned by the whole electorate of that particular constituency. *The Bhutiya Lepcha candidates had to be elected first by Bhutiya-Lepcha voters and thereafter to be voted again by the whole electorate of the constituency.*[42]

When election was first held on this basis in 1953 all the 6 Bhutiya Lepcha seats were won the National Party and the 6 Nepali seats by the Sikkim State Congress. *The voice of the 75% of Nepali population was reduced to a minority because the National Party along with nominated members of the Chogyal formed the majority in the State Council.*[43]

Parity System Modified

The parity system was slightly modified by the Royal Proclamation of 1958.[44] The number of Councillors was raised from 17 to 20 with 6 representatives of Nepali and 6 of Bhutia Lepchas. One seat was reserved for the *Sangha and one was declared general. There was a further change in the election law in 1966. The elective seats of the Nepalis and Lepcha Bhutiya were increased to 7 each. 2 new seats, 1 each in the name of Tsongs and schedule castes were introduced besides the usual one reserved for the Sangha, 1 general and 6 nominated seats. The total strength of the Council became 24.*

This was very much resented to by the political parties[45] other than the National Party. It was interpreted as an attempt to introduce casteism along with communalism and thereby cut into the Nepali hitherto considered as Nepalis who voted as one in all previous occasions. This system of parity, so much against the principle of democracy is still continuing with slight modifications as a unique feature of Sikkim's administrative system.

India had a direct role or so the Nepalis alleged in the introduction of this system. The Nepalis, who felt deeply wronged by this party system, accused India of getting it introduced in Sikkim with a definite objective of fuelling communal feeling in order to keep the two communities perpetually divided. *This would serve India's interest, they analyzed, by checking the growth of a feeling of pan Nepalism among the Nepalis spread from the hills of Darjeeling and Sikkim to Bhutan and Assam. This would keep the Nepalis in check, at least, in Sikkim.*

The Nepalis further grudged that the Indian officers had always sided with the Bhutiya Lepchas whenever there was any clash of interest between the two communities.[46] This principle of parity once introduced in election system did not remain confined to this alone. In course of time it extended to other fields as well like the award of stipends and scholarships and eventually the employment. *Seen from another angle, it may be said that the state, as a whole, continued represent the Buddhist culture, Buddhism was the state religion and being so the numerous Buddhist monasteries received state and annually for their maintenance.*

Subsidy for Cultural Events

The New Year Dance festival of the Buddhists and the worship of the snow ranges of kanchenjungha, also a Buddhist festival, received subsidy from the exchequer. This also created a certain sense of grievance among the Nepalis. On specific reason of their discontent was that even though their language was virtually the *Lingua frenca* of the country it was not accorded official recognition. The government also framed rules for the protection of the land rights of the indigenous population. Settlement of persons of communities other than Lepcha Bhutias was prohibited without express permission of the Government. Another instance of the anti Nepali policy of the Chogyal was the issue of citizenship in 1961.

Nepalese Excluded

In July 1961 the Sikkim Durbar promulgated the Sikkim subjects regulation providing the rules regarding citizenship of the various residents of Sikkim. It referred to Sikkimese, Bhutiyas, Lepchas and *Tsongs* among the categories of persons entitled to citizenship excluding the Nepalese. The Sikkim State Congress and Sikkim National Congress whose main base was the Nepali population strongly denounced the discrimination and brought it to the notice of the government of India. Finally on the latter's intervention, some changes were affected in it.[47] Thus, the political atmosphere of the state was totally vitiated with feelings of communalism and mutual between the Lepcha Bhutiyas and the Nepalis.

The Nepalis, with their demands for equal treatment and democratization of administration came to be identified with wheat went against the interest of the Chogyal and indirectly the Bhutiya Lepcha. One issue that created turmoil in the society and pointed to the discriminatory policies of the Chogyal in the premerger was the above mentioned system of parity—a blatant example of how the Bhutiya Lepcha, inspite of being a minority community was enjoying by virtue of the Royal patronage of position more privileged then the majority Nepalis. The Nepalis were extremely disgruntled tdishar equal number of seats with the Bhutiya-Lepchas. This even led them to rise in revolt against the Chogyal in 1973. After it, an agreement signed by the three parties-the Chogyal Government of India on

May 8 1973. It provided for an Assembly elected on adult franchise. The pattern of reservation was to be such that no single ethnic community would dominate over the other. To ensure this, party formula was again followed.

Hence, 15 seats were reserved for the Nepalis, 15 for the Bhutiya Lepchas, for Sangha and 1 for the Schedule Castes. The seat for Sangha was virtually for the Bhutiya Lepchas while the one for the Schedule Castes represented the Nepalis.[48] This pattern of reservation continued till 1979 when the Janata Government, under Morarji Desai scrapped it through an ordinance. The Bhutiyas and Lepchas henceforth came to be designated as Schedule Tribes. *The new system that substituted the party provided 12 seats for Schedule Tribes 2 for the Schedule Castes 1 for Sanga. The remaining 17 seats were general.*

The Major Issues Since 1975

Some basic changes were introduced in the administration of Sikkim in the post merger period. The monarchy was substituted by a democratically elected government and Buddhism ceased to be the state religion in Sikkim which now became a part of secular India. Even though the Bhutiya Lepchas continued to enjoy certain privileges by virtue of the special status accorded to Sikkim by the Indian Constitution after 1975, they no longer enjoyed the kind of patronage that they were enjoying previously.

Sikkim was opened up to the 'outsiders' i.e. form people from other parts of India and their migration in masses once again changed the demographic profile of the state. All these changes brought in certain socio-economic transformations and complexities in the state. The question of relation between the Bhutiya-Lepchas and the Nepalis also assumed a different form. The relationship became more pointed.

Increasing Nepali Participation in Sikkim Politics

Reservation of seats: A Demand of the Nepalis

However, one issue that still continued to dominate the political scene for the some more years was that of parity. Ironically, now it was the turn for the Nepalis to demand its restoration, even though they had condemned it so strongly before. It is also significant that the Nepalis, who greatly outnumber the Lepcha Bhutiyas should demand such a privilege in the electoral system of the state. It is precisely on this ground that the Centre shows reluctance to restore the system. The question, it points out, is why should there be seat reservation for the majority community.

It also agrees that restoration of such a privilege would automatically lead to a similar demand from the Nepalis settled in other parts of India. The Nepalis of Sikkim, in their turn point out the steady influx of outsiders from other part of India as a justification of their demand. Trade and business, they argue, has long been controlled by plainsmen from other states. Nepalese people from Darjeeling and Kalimpong, as distinct from the Nepalis of Sikkim, are also coming in.

Outflux

All this has created a fear among the local Nepalis that if this process of outside influx continues at the same rate, they might even lose their newly acquired political dominance. Since it was not possible to stop influx, they felt that reservation was the only way out to prevent such an eventuality.[49] The issue was as expected, a very strong one immediately after 1979. The then Chief Minister Mr N.B. Bhandari took it up and it even cost him his office. In a meeting of the Party in Jorethang, he put forward this demand along with the two other demands of citizenship and language. The Governor Mr. Homi J.H. Talyarkhan, with whom Bhandari's personal relation was already estranged and Mr. P. Venkatasubbaiah, the Union Minister of State for Home Affairs, asked Mr. Bhandari to give up this demand on reservation. The ultimatum was repeated and when Mr. Bhandari stubbornly stuck to his position he was finally dismissed. No reason was assigned.

Naturally, the issue proved to be a burning one in 1985 election and became the main plank of the campaign of Bhandari's party, the Sikkim Sangram Parishad. The issue was so strong that even the national party like Congress (I) was compelled to lend support, though in a less strident manner to this regional demand. Bhandari was returned to power with an overwhelming majority bagging 30 out of 32 seats. Since then, he has repeatedly pointed out that the old formula was suitable for the State and should have been restored.

According to him the system was conducive to communal harmony but even then it needed a review, because the businessmen who had already settled in the State be given representation. The idea being reservation, he said, was the need to maintain communal harmony in the border state. The formula of reservation that he suggested offered 13 seats to the Nepalis, 13 to the Bhutiya Lepchas 2 to the Schedule Castes, 2 to the Limbus and 2 to the others, meaning the non-Sikkimese.[50] In 1986,

Bhandari gave one year time to the Centre for fulfilling the demand and in 1988 he threatened to launch an indefinite fast to realize his demand. He even presented memorandum to the Prime Minister following which he was given assurance that the issue would be discussed among his cabinet colleagues with a view to solution. His fast, however, was put off following a request from Delhi.[51]

Finally, he appointed a high level 7 members committee in 1988 to look into the viability of the demand. The Committee included highly placed officials like the Chief Secretary. Home Secretary and the Advocate General of the State. The Committee, in its report observed that Nepalis from other parts of India, who are actually Indians of Nepali origin were not the same as Sikkimese Nepalis. These two groups, the report pointed out, were discernible by clear and identifiable criterion.

The Committee further pointed out that the amendment to the Representation of the People's Act of 1980, modifying the reservations of seats for the different communities was not totally contrary to the Tripartite Agreement of May 8, 1973. The Amendment was alleged to have abrogated the reservation for the Nepalis of Sikkimese origins. This was viewed as detrimental to the social, economic and political interests of Sikkim.[52] Bhandari stuck to his demand for restoration of reservation of seats for the Nepalis till as late as 1991. He pointed out that the abolition of seat reservation went against the spirit and letter of the merger of the State with India and was a great betrayal and injustice to the people of Sikkim. He, however, ruled out the possibility of any ethnic violence on this ground.[53] In short, theis demand for reservation of seats for the majority community is a peculiarity of Sikkim politics, a result of Sikkim's particular status and demographic composition.

However, this may not be able to satisfy all for good. The possibility remains that it might lead the different group and communities of people, speaking the same language and so far denoted by the broad generic term 'Nepali' to come up with new demands for further reservation of seats for themselves or delimilation of constituencies. The issue is taken up every now and then out of political expediency by all the concerned parties who come up with new formulae of reservation of seats for different sections of the population.[54]

This demand for restoration of reservation of seats for the Nepalis of Sikkim is aimed mainly to serve the interest of the Nepalis inhabiting the State for long. The two other demands, viz. the demand for grant of citizenship to the stateless Sikkim subjects and the recognition of Nepali in the Eight Schedule of the Constitution too, were aimed at serving the interest of the same community. The citizenship issue too, like the one relating to reservation cropped up every now and then in Sikkim. The problem lay in the disagreement between the Centre and the Sikkim Government over the issue of immigrant Nepalis.

Citizenship

The Centre wanted to solve the issue on the basis of the Sikkim Subject Regulation of 1961, formulated on recommendations of the late Chogyal. The Chief Minister, on the other hand, argued that no addition was made in the list of resident citizens when Sikkim was merged with the Indian Union in 1975. As a result of this, several thousand persons of Nepali origin had become outsider in the land of their birth. The Central Government did not accept this line of argument. According to one highly placed Union Home Ministry official, soon after the formal merger of Sikkim, the Union Home Ministry, in a notification dated May 6, 1975 had extended section 7 of the Citizenship Act of 1955, thereby providing enormous scope for the people of Sikkim to forward their claims for grant of citizenship should it have been denied to them. It is rather untenable he said, that these people should keep silent about it for more than 10 years and then it should suddenly dawn on Mr. Bhandari that several thousands of Indian Nepalis in Sikkim had been denied citizenship.

After years of negotiation, the Sikkim citizenship issue was finally resolved in August, 1990 when the Government decided to grant citizenship to 75,000 stateless people living in Sikkim for several decades. It was revealed that cases of another 5000 such people were also under consideration of the Home Ministry.[55]

It was around this time that a romour was afoot regarding plans of a merger of Sikkim with Darjeeling district. Simultaneously with it went the rumour that Bhandari was desirous of becoming the leader of the Nepalis in India. Not only was he taking interest in the GNLF Movement in Darjeeling and campaigning the cause of the inclusion of Nepali language in the Eight schedule,[56] as early as 1975, he had invited all the Nepalies of Assam to come to Sikkim if Assam refused to have them. This speech, given in North Lakhimpur, Assam, was alleged to have led to the burning of hundreds of houses in communal clashes on the banks of Brahmaputra.[57] There was also the memory of long and close association between Mr. Bhandari and Mr. Ghising since the Gorkha League days.

Impact of the Bhutiya Lepchas

Tribal Grievance Over the Election System

The increasingly Nepali tilth in Sikkim politics under Mr. Bhandari had its repurcussion on the Bhutiya and the Lepchas, who it was felt, were finding themselves increasingly alienated from the political mainstream. The fear of cultural extinction and nostalgia over lost political dominance brought the tribals closer to each other. Several organizations have been formed with a view to campaigning actively for tribal unity and give vent to their grievances regarding the electoral system and the administration of the state. With respect to the election system, the main grudge of the tribals is the demographic composition of the constituencies, which make them dependent on the non tribals. They point out that even though 13 out of 32 seat of the State Assembly are reserved for them, they have to depend on the votes of the majority community for getting elected in all but the 3 reserved constituencies in the North. This is because majority of the voters in the reserved constituencies are non-tribals.

Voters

Figures regarding the number of voters would explain the problem. The total number of voters for 32 seats of the Assembly was 105,000 in 1979 which rose to 150,000 in 1985. Even though a study of the final list indicated that tribal voters had increased by about 22.56 per cent a constituency wise study constituencies–even in the North, which used to be predominated by the tribals. According to a breakup of the voters community wise, it became clear that in Dzongu there has been a 100 per cent increase in the number of all Nepali voters. Similarly, in Lachen Mongshilia constituency the increase has been by several hundreds–from just 84 to 864.[58]

The Problem was brought to the notice of the Prime Minister by the Sikkim Tribal Welfare Association, (the first of its kind and formal immediately after the merger with India) through a memorandum. In it, it was particularly pointed out that in North Sikkim, where the Bhutiya Lepchas used to be dominant, gradual influx of Nepalis has started threatening their existence and culture.[59] It is difficult to suggest a solution to the problem a the tribal population is scattered allover the sate. The formula that the tribals thought could serve their interest was the system of adult franchise practiced in the Chogyal Government.

However, the system, also known as confidence vote had to be done away with due to its complicated nature and could not be introduced anew.[60]

Reorganization

The other formula that the tirbals approved was the reorganization and delimilation of the constituencies within the district in such a manner that the tribals would cast their vote in favour of the tribals alone. In this connection, the demand for an Autonomous Regional Council for North Sikkim has also come up from time to time.[61]

Second organization, the Sikkim Tribal Yargay Chogpa too had been actively campaigning for tribal unity. At its first state level convention the Chogpa called for unity among Bhutiya, Lepchas and the Limbus to strengthen their hold on affairs of the state. It also demanded tribal statues as well as reservation of 6 states in the State Assembly for the Limbus. This number would be at per with the seats already kept reserved for the Bhutiya and the Lepchas.[62] The same feeling of insecurity experienced by the Bhutiya Lepchas was given vent to by Thugchuk

Lachungpa, general secretary of the Sikkim Tribal Welfare Association in a press conference in November 1988. Mr. Lachungpa, the then Forest Minister of Sikkim, pointed out that under the existing system the tribals have to depend on large Nepali vote banks to get elected.[63]

Interest of the Ethnic Communities

The Sikkim Tribal Yargay Chogpa which rivals the Sikkim Tribal Welfare Association even gave a threat to launch a mass agitation for safeguarding the interest of the ethnic communities in the state. Its General Secretary, Mr. Dorji was more vocal about the problems arising out of the large scale influx of outsiders into the state. Due to this, not only the Bhutiya and the Lepchas but even the other backward communities like Gurung, Rai, Mangar, Tamang and Bhujel settled in Sikkim for years had turned into minority communities. He even described Bhandari's demand of the stateless Nepalis as unjustified. All of them were Nepali people, he pointed out.[64]

A 5 members delegation of thee tribal people was sent to Delhi to familiarize the Union Home Minister, Mr. Buta Singh and some other cabinet members about the problem. The delegation told them that the only way to safeguard the interest of the SCs and STs was to set up an autonomous district council in terms of the 6th schedule of the Constitution. The other suggestions related to reservation of an additional Parliamentary constituency for the STs and same additional stipends and facilities for student belonging to these communities.[65] Another organization supposedly representing the interest of the Bhutiya Lepchas Denzong Lho Mon Chodrul was formed as lat as August 1992. One main demand of the organization has been the restoration of the 15 reserved seats and the reservation of an additional seat for tribals other than the Bhutiya and the Lepchas.

Reduction of Reserved Seats

The reduction of reserved seats from 15 to 12, in their opinion, was a violation of the terms of Morger. The Chodrul, in its memorandum to the Chief Minister has also mentioned the need to retain one reserved seat for the Buddhist Lama in the State Assembly. It alleged that certain elements in the State were trying to reduce the number of seats reserved for the Bhutiya Lepchas as well as to strike out the one Sangha seat. The political rights, religious sentiments and age old land laws, it argued, needed to be protected so that the fear in their minds was erased for ever.[66]

Land: A Second Issue of Tribal Apprehension

A second issue that has disturbed the tribals for long has been that of land. This point needs to be elaborated only in the context of a short description of the land holding system as it used to be in Sikkim.[67]

Landholding System: As it Used to be

Originally, before the arrival of the Bhutiya, the Lepchas were owners of the land and later, after the installation of the Namgyal Dynasty the Bhutiya king became the owner of the same. The king used to gift portions of land to his courtiers who were mainly kazis of either Bhutiya or Lepchas origin. Some changes were introduced by J.C. White, the first Political Officer which led to the emergence of some early Nepali settlers, particularly the Newars as lessee landlords, later designated as Thikadars. This system of lessee landlordism continued till the implementation of 1950 land reform programme after which the lessee system was ebolished.

As already mentioned in the foregoing sections, the Chogyals followed discriminatory policies in matters of land revenue administration. Here follows a short description of some of these discriminations. Land was distributed to the Kazi families for 15 years while it was given to the Nepali thikadars for 10 years. A law was introduced whereby alienation of land by hereditary state subjects i.e. the Bhutiya Lepchas in favour of non-hereditary subjects i.e. the Nepalis was prohibited. The Sikkimese Bhutiya Lepchas also enjoyed the right of exclusive acquisition of land and residence in North Sikkim. There was discrimination also in terms of assessment.

Discrimination against Nepalis

The Nepali cultivators were assessed at a high rate than the Bhutiya Lepchas. This was brought to an end only in 1966 when the rate of which the Nepalis were assessed was imposed upon the Bhutiya Lepchas as well. Discrimination was practiced in the land ceiling as well. The Kazis, the Thikadars and Mondals hold unaccountable amounts of land earlier and they still own huge plots of land. In 1974, a land ceiling was imposed according to which, the *bustiwallas* could not hold more than 20 acres of land, mondals not above 30 acres and Kazis and Thikadars not above 100 acres. The last category of people, however, were allowed to hold land up to 200 acres with Government's approval and subject to the payment of Rs. 5 per 20 acres excess. However, there was no limitation imposed on their common fields, orange garden, homestead, fodder and fuel forest etc.

The settlement report of 1958 pointed out that the highest per centage of total cultivated land amounting to 66 per cent was owned by the Nepalis while the Bhutiya Lepchas owned 20 per cent and 14 per cent respectively. It was the Bhutiya population that had the highest amount of excess land.

According to a recent survey conducted by the Land Revenue Department of the Government of Sikkim, the amount of excess land (from the ceiling) fixed on the basis of community was as follows:

Bhutiyas	-	2795.01 acres
Lepchas	-	645.40 acres
Nepalis	-	1352.83 acres

However, acquisition of such land has been reported to have succeeded only partially, those who carry out the duties being more than not, those who own excess lands or those who are otherwise affiliated to the person having excess land.

Post Merger Legislations

A number of land reform legislations have been enacted in Sikkim since 1975. The most important of these are. (1) The Sikkim Cultivators' Protection Act 1975, (2) The Sikkim Agricultural Land Ceiling and Reforms Act 1978 and (3) Sikkim Land Requisition and Acquisition Act, 1978.

The main objective of the legislations are three fold :

1. Elimination of intermediaries in the land management.

2. Enforcement of ceiling law in respect of land holdings and distribution of surplus land and

3. Protection of tenancy rights. Keeping with the tune of these objectives the Sikkim Transfer of Land Regulation Bill of 1989 was introduced. This is commonly known as the Bill No. 5 to 1989.

Tribal Reaction to the Legislations

Based as these laws are on the democratic concept of limiting the size of holdings and protecting the tenancy rights, they have become issues of great concern for the Bhutiya Lepchas who have traditionally been big land owners. Very recently, the tribal organization viz. Denzong Lhomen Chodu/Rong Arat Sa Tarzum in a resolution of their conference pointed out and explained at length the clauses that have disturbed them the most.[68]

According to this resolution, it is the Sikkim Land Ceiling and Reforms Act 1978 that hs disturbed them the most. In it they are much concerned about the clauses like :

1. Vesting of excess land above the ceiling limit with the government and

2. Distribution of vested lands to those persons who are eligible for grant of such land. Not only they feared partial loss of their holding but were all the more concerned because the allottees, they feared, were going to be non-tirbals. Provisions of section 28 under Chapter VII of the Act, they maintain, clearly indicated that the Bhutiya Lepchas tribals were going to be imposed with land ceiling.

Sikkim Transfer of Land Regulation Bill 1989

The Sikkim Transfer of Land Regulation Bill 1989 is the second cause of their grievance and apprehension. The law not only elaborates Revenue order No. 1 of 1917 but at the some time undoes some of the laws analogous thereto and incorporates certain provisions, which they think are vague and as such prone to misuse or abuse.

The first problem for them arises out of the term 'Competent Authority'[69] that has been given an overriding power to allow alienation of land belonging to Bhutiya or Lepchas communities of Sikkimese origin in favour of other communities. What the tribals find objectionable is that the competent authority has not been specified. The state government has been given unfettered power to appoint nay person as such authority. Under such circumstances, they are afraid the powers so vested may be misused against them. Hence their demand that this competent authority should rather be defined in specific terms. In fact, the issue of land has figured on the agenda of programmes of other previous tribal organization s as well viz. Denzong Yargay Chogpa (1987-88), Joint Action Committee of the Bhutiya Lepchas Tribals, Lho Mon Youth Council, etc.

Land of Unity

. However, despite formation of organizations they have not really been able to achieve much in this respect. One important reason of this is the lack of unity and commitment among themselves. Mr. P.O. Pazo, a Lepcha himself, who had acted for years as a president of the above mentioned Sikkim Tribal Welfare Association himself admitted that neither he nor his predecessors Mr. Ranzing Dorjee or Mr. Sonam Dorjee had been able to achieve much. He gives the reason as that he himself was a government office and being so, could ill afford to go to extremes with his pro-tribal issues against the government.

According to his statement, the founding members of the organization were removed following manipulations by the government. The latter placed its own nominees at the various offices. Being pro government, they did not do anything that would go against the interest of the government.[70] The issue of land had come in the forefront when land was gifted to N.B. Khatiwada by L.D. Kazi. This transfer of tribal land to a non-tribal brought the tribals together in their attempt to protect their interest The Joint Action Committee of the Bhutiya Lepchas Tribals was constituted-Once again headed by M. Pazo. They even moved the Supreme Court to defend their right. Mr. Pazo formed a third organization, the Lho Mon Chodrul Rong Arat Sa Tarzum after the Bill No. 5 of 1989 was introduced in the Assembly. However, the only success that he claims to have achieved till now is that the government has been a bit cautious in the implementation of the law.[71]

Emerging Right between the Bhutiya Lepchas

From what has been discussed so far, it appears that it is the religion that has always been the basis of polarization in politics. In other words the alignment between the Buddhist Bhutiya and Lepchas has been so strong that they have always acted as one single community against the Hindu Nepalis. These Bhutiya and Lepchas have now been accorded the status of tribals and the problems of ethnic relation is generally referred to as one between the tribals and the non tribals. But in reality this apparently cordial relationship between the Bhutiya and the Lepchas does hide an under-current of tension, which become apparent only at certain points of time. The main grievance on part of the Lepchas is that the Bhutiyas, being economically stronger than them, have always stolen the leadership of the tribals and in the name of tribal unity underplayed the interest of the Lepchas.

An Early Case of Lepcha Revolt

The first banner of Lepcha Revolt raised by one Ruth Karthak Lepchani was directed against the Sikkim Durbar.[72] On the eve of 1967 elections to the Sikkim council Ruth launched one party named Sikkim Independent Front and fielded 6 candidates, 5 of whom were Lechpas and 1 was Nepali. The nomination papers were rejected and she herself was arrested and later sentenced to two years vigorous imprisonment in November 1968. The government of Sikkim, however, released her in March 1969 with orders to quit Sikkim. She hurriedly left Sikkim and crossed over to Kalimpong.

Since we do not have any document to reveal her point of view, her statements can only be reconstructed form the complaint failed by the Chief Secretary against her. In it, she was accused of inciting the Lepchas by saying that the real rulers of Sikkim were the Lepchas and the pro Bhutiya policy of the Chogyals,—who actually had no right to rule over Sikkim, was only suppressing the Lepcha whose socio-economic condition was steadily deteriorating. This was naturally interpreted as defiance of the authority of the Chogyal and accusing him of being partisan. As this would creates discontent among the Lepchas and incite them to rise up against the Monarchy, this could not be allowed to continue Ruth was accused of challenging the very foundation of the state. Since we know little about this lady Ruth Karthak Lepchani, it is difficult to explain her actual objectives and motivations. No voice of protest was raised against the whole things. Even though there was a certain Lepcha Association that enlisted members and raised funds for the protection of the right of the Lepchas, they maintained a silence over the issue.

Similar welfare associations for the Lepchas have been there in Sikkim and their grievance against the Bhutiyas even if it existed was not expressed openly.

Some Recent Developments

It is only recently that the undercurrent of tension between the two communities have come to the surface and voices raised against Bhutiya predominance over the tribals as a whole. Some Associations to protect and further the right of the Lepchas have been formed. One of them is Renjyoung Mutanchi Rong Ong Shejum of Gangtok. It was established in 1992.[73] In a direct representation to the Prime Minister of India, they have expressed a feeling of being left out from the national mainstream. Though there are many cases behind their such sentiment, the primary reason seems to be a grudge against the Bhutiyas, who, they complain, are fast sideling the Lepchas.[74]

In their representation, they point out that this attempt on part of the Bhutiyas is reflected in a recent representation sent by 'Denzong Lhomen Chodu' to the Government of India seeking amendment of the Constitution (Sikkim) Schedule Tribes Order 1978, Representation of People's Act 1950 (Act VIII of 1980) and also Act 1951 (Act VIII of 1980). In the schedule of the Constitution (Sikkim) Scheduled Tribe Order 1978, the names of the tribes were thus states–

1. Bhutiya (including Chumbipa, Dopthopa, Dukpa, Kagatay, Sherpa, Tibetan, Thromopa Yolmo) 2. Lepcha.

The Denzong Lhomen Chodu has sought the amendment in the following manner.

1. Bhutiya. 2. Lepcha. 3. Chumbipa.

4. Dopthopa. 5. Dukpa. 6. Sherpa.

7. Tibetan. 8. Yolmo.

The representation goes on to argue that this suggested arrangement is incongruous. All the tribes mentioned before and after Lepcha belong to the same stock of Bhutia. The Lepchas, on the other hand, "being the indigenous tribes of this area have their own culture, language, script and ancient historical background". Hence their protest against being sandwiched between tirbes of the same stock in this suggested amendment in the schedule. Their second grievance is their backward state of the development in comparison to the Bhutiyas. Their concentration in the far flung areas of Sikkim and economically backward state have turned them into a marginal class in the State of Sikkim. Hence their demand for a 50 per cent reservation of all tribal benefits in both educational institutes and jobs for the Lepchas.[75]

Land Revenue Legislation

As to the land revenue legislation of the Government of Sikkim, the Lepchas have their own grievance. While they fully agree with the Bhutiyas in their protest against the land revenue legislations of 1978 and 1989, they point out how the Bhutiyas themselves, remaining within the limits of law are exploiting the Lepchas. As per the records in the Land Revenue Department, the Lepcha land seems to have passed into easy ownership of other tribes while there is hardly any record of any Lepcha having bought any land from others. If at all they have it is the rerest of things. There are many who often takes a Lepcha land for a paltry amount. The hint is obviously towards the economically stronger Bhutia community. The Lepcha demand, therefore is for a legislation to confine the transfer of Lepcha land within the Lepcha community alone.[76]

Among their other demands there is the one for a policy for the development and diffusion of their language through permanent post of news readier in AIR Gangtok and introduction of their language at the degree level, etc.

Conclusion

The Bhutiyas and the Lepchas, by virtue of their previous position, are sitll enjoying certain privileges in the society. Reservation of 12 Assembly seats itself is an example of this. Coupled with the one reserved seat for the Sangha, this make up more than 40% of the total number of seats, while, they as a group now constitute les then 25% of the population. The reservation of one seat for Sangha is also a mark of recognition of the particular course of Sikkimese history, where religion and monasteries played an important role in the administration of the state.

The recent legislation regarding income tax also comes to serve the interest of the tribals.[77] The tribes, particularly the Bhutiyas are economically well off in comparison to the common Nepali masses. Yet it is they who are exempted from the payment of tax. In fact, this has been a major issue in recent Sikkim politics that has only depended the rift between the tribes and the Nepalis. However, realizing the rationality of Nepali dissatisfaction over the issue the Sikkum Tirbal Welfare Association has urged the Centre to extend the benefit to the Nepalis of Sikkim as well and bring them at par in the tirbals.[78]

Yet, these concessions, the tribals feel, are no match to their loss of kingdom and their feeling of deprivation and being 'wronged'. In fact, these are exactly the feelings that have gone a long way to shaping their attitude to the Nepalis in recent times also. The abundance of tribal organizations representing the interests of the tirbals point out their unhappiness with the state of affairs in Sikkim. Ironically it is the abundance of such organizations that also bring out the weakness of the tribals–a lack of unity and cohesion among them. Organizations are formed every now and then as and when an issue crops up but there is an inability to pursue the goals in a concerted manner.

Even now, there are two organizations (among many others) that profess two extreme views. On the hand, there is the organization named Survival Sikkim that claims that Sikkim exists only for Bhutiyas and Lepchas and the rest are neo colonists. Recently, it has condemned Nepalis on the eve of Bhanubhakta celebration–Bhanubhakta being a symbol of unity among the Nepalis.[79] On the other hand, there is also organization called Nabhola that claims to represent all the three Bhutiya, Lepchas and Nepali communities, the main objective of it being the promotion of peace.[80] Under the circumstances, the tribals are not in a position to achieve much, excepting some concessions here and there. The age old mistrust and the apathy towards the Nepalis continue with the difference that while prior to merger with India they were dictating terms to the Nepalis from a position of authority, now they are on the defensive, fighting out of a feeling of insecurity for grant of some special privileges that would maintain their identity.

Perhaps, their consolation lies in the fact that, though reduced to a quarter of the population, they still constitute a considerable minority on whom even the Nepalis depend for the survival of their government. This is exactly what happened in 1995 when with 6 tribal legislators deciding to abandon Mr. N.B. Bhandari's Sikkim Sangram Prishad and join the ruling Sikkim Democratic Front, P.K. Chamling, the then Chief Minister of Sikkim, was able to consolidate his political strength.[81]

References

1. H.H. Risley in '*History of Sikkim* an its Rulers' in *The Gazetteer of Sikkim*, first published in 1894, Reprint in 1985, by B.R. Publishing Corporation, Delhi, p. 27.
2. *Ibid.*, pp. 34-36. Risley gives the names of these families and points out that many of these families give mythical origin to their ancestors and trace their descent from spirits, demons or from an alliance between a rock, a tree, or a pond. To these they assign a known location.
3. A.C. Sinha, *Politics of Sikkim; A Sociological Study*, Thomson Press (India) Limited, Publication Division, 1975, p. 12. His sources are various works of Col. B.G. Mainwaring, K.K. Das and the *History of Sikkim* by T. Namgyal and Joshay Dolma, MSS, 1908.
4. *Ibid.*, p. 7.
5. H.H. Risley, *op. cit.*, pp. 36-37.
6. *Ibid.*, p. 37. There is a controversy among scholars over the question of the origin of the Limbus. T.B. Subba discusses them in brief in article 'Limbu ethnicity at cross roads' in Himalayan Miscellany, Vol. 4, 1990, pp. 40-41.
7. H.H. Risley, *op. cit.*, pp. 28-31. Risley names all the families and their subdivisions and also enumerates the legends connected with their origin. In this context it may be stated that two families Lbar-Phong Pu Tsha-po (Barphungpuso) were hitherto thought to be pure Lepcha or Rong families. General Mainhearing considers that the fist is Bar fong mo = "the flowing from the high" and the second Aden "the created, fashioned, formed"–the two combined

being equivalent to 'the patricians and plebians'. Risley however, maintains that the accounts given by the members of the two families are different. By analyzing their history and genealogical table he shows that their descent is not at all a pure Lepcha one and in these two families there were marriages to Limbu, Mangar and even Royal Bhutia families.

8. A.C. Sinha, *op. cit.*, pp. 8-9.
9. H.H. Risley, *op. cit.*, p. 27.
10. J.C. White, 'Sikkim and Bhutan: Twenty-one years on the North East Fronteris 1887-1909, Reprint 1986 by Manas Publications, Delhi, pp. 7-9.
11. Urmila Phadnis, in her book Ethnicity and National Building in South Asia, Sage Publications, New Delhi, 1990, p. 14 defined an ethnic group "as a historically formed aggregate of people having a real or imaginary association with a specified territory, a shared cluster of beliefs and values connoting its distinctiveness in relation to similar groups and recognized as such by others". For an analysis of the definition and divergent perspective and approaches to ethnicity see pp. 13-20. Also see Paul R. Brass, Ethnicity and Nationalism: Theory and Comparison, Sage Publications, New Delhi, 1991, pp. 18-23.
12. *History of Sikkim, op. cit.*, pp. 44-45.
13. *Ibid.*, p. 76.
14. *Ibid.*, p. 50.
15. H.H. Risley also refers to this while analyzing the genealogical tree of he Barphungpuso family of his Rui Chhung, to which belonged Tasha Apong. See. Ff.7.
16. *History of Sikkim*, p. 122.
17. *Ibid.*, p. 128.
18. *Ibid.*, p. 129

19. A detailed description of the whole incident in *Ibid.*, pp. 138-39.
20. *Ibid.*, p. 147.
21. Chie Nakane: 'A Plural society in Sikkim: A Study of the International of Lepchas, Bhotias and Nepalis' in Caste and Kin in Nepal, India and Ceylon, Anthropological tudies in Hindu Buddhist Contact zones ed. by Christoph Von Fu rer Haimendorf, Asir Publishing House, 1966, p. 220.
22. *Ibid.*, p. 221.
23. Limbu Kingdom that existed in Easter Nepal until 1764. Limbuana fell into the hands of Gorkha king Prithvinarayan Shah.
24. A.C. Sinha, *op. cit.*, p. 14.
25. *Ibid.*, pp. 14-15.
26. *History of Sikkim*, pp. 148-49.
27. H.H. Risley, *op. cit.*, p. 10.
28. A.C. Sinha *op. cit.*, p. 15.
29. *Ibid.*, p. 14. T.B. Subba *op. cit.*, p. 43.
30. The Association continued till some 20 years back but it could never really be very active. T.B. Subba gives various reasons of its failure.
31. *History of Sikkim*, p. 76.
32. J.C. White, op., cit., p. 9.
33. This refers to the Bhutanese and Gurkha inroads into Sikkim during the reign of Tenzing nemoue (1769-1790). The Bhutanese attack took place in 1770. They overran all Sikkim, cast of Tista but were ultimately thwarted by Sikkimese troops. The Gurkha attack took place in phases from 1776 onwards. The final attack was in 1788-89 when they overran the whole of Sikkim south and west of Tista. The king and the Queen fled Sikkim and in 1790 reached Lhasa to obtain help. The King died there in the same year.

34. *History of Sikkim*, p. 68.

35. *Ibid.*, p. 69.

36. *Ibid.*, p. 75.

37. *Ibid.*, p. 77. A detailed description of the course of events regarding the Nepalis settlement and texts of the letters thus exchanged between the Durbar and the Bengal government in para 68-78. For the political developments of the period and Gurkha immigration also see L.B. Basnet, A Political *History of Sikkim*, S. Chand & Co. (Pvt) Ltd. New Delhi, 1974, pp. 18-50. Also in Sunanda K. Datta Roy smash and Grab: Annexation of Sikkim. Vikas Publishing House, New Delhi, 1984, pp. 37-38.

38. Initially, 3 parties viz. Praja Sudharak Samaj, Praja Sammelan and Praja Mondal were formed to voice protest against landlordism and feudalism. These parties merged together to form the Sikkim State Congress on 7.12.1948. For details regarding the political developments of this period see L.B. Basnet. *Op. cit.*, pp. 82-88.

39. *Ibid.*, pp. 94-95. a full text of the treaty in pp. 204-208. S.K. Datta Roy also discusses the nature of the treaty and the Indo Sikkim relationship after this in his above mentioned book.

40. L.B. Besnet *op. cit.*, pp. 97-98.

41. *Ibid.*, p. 99.

42. *Ibid.*, pp. 99-102, see foot no. 59.

43. *Ibid.*, p. 102.

44. *Ibid.*, p. 108.

45. *Ibid.*, p. 134.

46. *Ibid.*, p. 172. This view is repeated again and again by Besnet in his bock. Basnet, himself an activist was active throughout the perioid as an anti monarchy politician. Not all other writers contribute to this oint of view Sunanda K.Datta Roy, for

example, in course of his discussion about how the Chogyal's relations with the Indian political officials worsened in course of the 1960s makes the specific statement "Durbar wa more convinced that it was Indian policy to weaken Tiboto Buddhist enthnic by extending support to Nepalese Hindus who might be expected to form alliances with their compatriots in Darjeeling". Sunanda K. Datta Ray, *op. cit.*, p. 71.

47. L.B. Basnet, *op. cit.*, p. 120.
48. *Ibid.*, pp. 179-188.
49. For a detailed discussion about the problem created the influx of outsiders see Karma Topden, "Sikkim: The Identity Question' in Himalaya Today, June-August, 1989, pp. 27-29.
50. *Times of India*, 3.3.85 and Hindustan Time 11.7.83.
51. The Statement, 4.11.88.
52. *Amrita Bazar Patrika*, 1.1.89.
53. The *Telegraph*, 29.5.90. The Statement, 30.5.90, 28.6-91.
54. The formula offered by K.C. Pradhan, an ex-minister of the Kazi Government may be cited inthis respect. He has classified the people of Sikkim into 4 categories:

    1. STs-Bhutiya Lepcha, Sherpa and Sangha.

    2. SCs–Kami, Damai, Sarki, Majhi.

    3. Backward Nepalis–Tsong/Limbu, Gurung, Mangar, Rai and Tamang.

    4. Nepalis of General category: Bahun, Chhetri, Newar. Pradhan.

    As per his formula, 6 seats each are to be kept for Bhutiyas and Lepchas, 1 each for the *Sangha* and the Sherpas. 2 seats for t CSs, 3 for Tsong and Rais, 1 each for Mangars and Tamangs, 2 for Gurungs and finally, 6 seats are to be allotted to the fourth category of the Nepalis in General. The Statement 15.1.89. Even recently N.B. Bhandari is reported to have claimed a 10% reservation of seats for the Nepalis in the State Assembly *Amrita Bazar Patrika* 25.5.95.

55. *Shillong Times*, 26.8.88, *Times of India*, 7.8.90.

56. Recognition of Nepali has been one of the major issues in the politics and movements of Darjeeling since the late 50s.The demand was initially raised by the Gorkha League and other associations and later became one of the focal points of the GNLF agitation in Darjeeling. Sikkim state Assembly passed a resolution favouring the issue in October 1977. For details of the language issue in the politics of Darjeeling see Dyutish Chakraborty Gorkhaland: Evolution of Politics of Segregation, Special lecture x of Centre for Himalayan Studies, North Bengal University pp. 26-32.

57. *Shillong Times, op. cit.*

58. *The Statesman*, 28.2.85. The number of voters is a little different in a report of the Indian Express 25.7.84. In it, the total number of voters was 117000 in 1979. As per the report of The Statement in 1985 75% of the voters were Nepalis, 20% Bhuitya Lepchas and of the rest about 5% were either plainsmen or from other communities.

59. *Amrita Bazar Patrika*, 11.12.88.

60. In this system, the winning candidates, besides obtaining the majority votes of their own community, had to obtain 15% of the total votes polled by the other community. If a candidate even after obtaining majority vote of his community failed to get 15% of the total votes polled by the other community, he or she would fail to get elected. The system was thought to be suitable for maintaining communal harmony by the then government of Sikkim but the whole procedure was complicated in nature.

61. For detailes, see Karma Topden, *op. cit.*
62. *The Statesman*, 8.11.88.
63. *Amrita Bazar Patrika*, 27.11.88.
64. *Ibid.*, 3.12.88.
65. *Hindu*, 5.4.89.
66. *The Statesman*, 4.8.92.
67. Discussion about traditional land holding system in Sikkim and the post merger legislations are available in several books and articles. Some of the papers consulted for this article are T.B. Subba. 1. Agrarian social structure and change in Sikkim' in Social change, March 1989 Vol. 19. No. 1.

    2. 'Agrarian Relations in Sikkim. The limits of change', paper presented at the seminar on Land. Land Relations and Land Reforms in the plains and valleys of North East India (April 1984) NE ICSSR, Shillong.

    3. 'Agrarian Relations and Development in Sikkim' in M.S. Mishra ed. Rural Development in Easter and North Eastern India, Calcutta, IIM, 1988. Amal Datta 'Land and Ethnicity in Sikkim' in Man in India 1992 72 (2) and the "Report of Sikkim settlement" by M.S. Bhandari UPCD (Retd). Settlement officer Sikkim 1958.
68. Memorandum submitted by Denzong Lhomen Chodu/Rong Arat Sa Tarzum, Ravangla, South Sikkim (Regd.) 591/LR/93-94 Vol. No. 1.
69. *Ibid.*, p. 5.
70. Based on a personal discussion with Mr. P.O. Pazo in Gangtok Sikkim in March, 1994.
71. *Ibid.*
72. For details, see L.B. Basnet, *op. cit.*, pp. 148-151.
73. Renjyoung Mutanchi Rong Ong Shajum (Sikkim Lepcha Youth Association) Diesel Power House Complex, Gangtok, Red. No. 553/LR Estd. 1992.

74. Copy of the memorandum pp. 1-3.
75. *Ibid.*, p. 3
76. *Ibid.*, p. 4.
77. The Financial Bill of 1994. It was to take effect from 1.4.95 and apply from the assessment year 1995-96 onwards.
78. *The Statesman*, 14.3.94.
79. *Telegraph*, 13.8.96.
80. *Ibid.*
81. For some discussion about the role of the Bhutiya Lepchas in State Politics see, *Amrita Bazar Patrika*, 18.5.94 and *The Statesman* 26.8.95.

Index